Unit Treatment Processes in Water and Wastewater Engineering

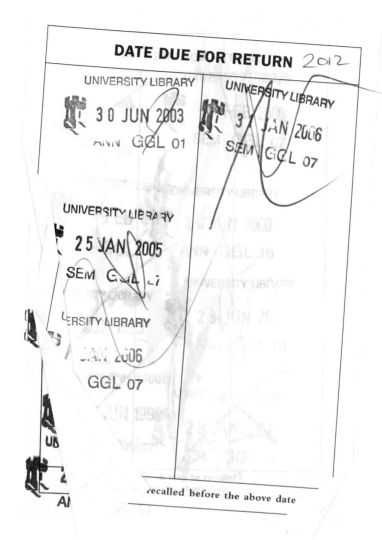

Wiley Series in Water Resources Engineering

The 1992 'International Conference on Water and the Environment: Issues for the 21st Century' served as a timely reminder that fresh water is a limited resource which has an economic value. Its conservation and more effective use becomes a prerequisite for sustainable human development.

With this in mind, the aim of this series is to provide technologists engaged in water resources development with modern texts on various key aspects of this very broad discipline.

Professor J. R. Rydzewski
Irrigation Engineering
Civil Engineering Department
University of Southampton
Highfield
SOUTHAMPTON
S09 5NH
UK

Design of Diversion Weirs: Small Scale Irrigation in Hot Climates
Rozgar Baban

Unit Treatment Processes in Water and Wastewater Engineering
T. J. Casey

Water Wells: Implementation, Maintenance and Restoration
M. Detay

Unit Treatment Processes in Water and Wastewater Engineering

T. J. Casey
University College Dublin,
Ireland

JOHN WILEY & SONS
Chichester · New York · Brisbane · Toronto · Singapore

Copyright © 1997 by John Wiley & Sons Ltd,
Baffins Lane, Chichester,
West Sussex PO19 IUD, England

National   01243 779777
International (+44) 1243 779777

e-mail (for orders and customer service enquiries): cs-books@wiley.co.uk

Visit our Home Page on http://www.wiley.co.uk
or
http://www.wiley.com

Other Wiley Editorial Offices

John Wiley & Sons, Inc., 605 Third Avenue,
New York, NY 10158-0012, USA

Jacaranda Wiley Ltd, 33 Park Road, Milton,
Queensland 4064, Australia

John Wiley & Sons (Canada) Ltd, 22 Worcester Road,
Rexdale Ontario M9W ILI, Canada

John Wiley & Sons (Asia) Pte Ltd, 2 Clementi Loop #02-01,
Jin Xing Distripark, Singapore 0512

Library of Congress Cataloging-in-Publication Data

Casey, T. J. (Thomas Joseph), 1993–
    Unit treatment processes in water and wastewater engineering /
T. J. Casey.
        p.   cm. — (Wiley series in water resources engineering)
    Includes bibliographical references and index.
    ISBN 0 471 96693 2
    1. Water — Purification.   2. Sewage — Purification.   I. Title.
II. Series.
TD430.C37   1996
628.1'62 — dc20   96–15921
                                                      CIP

British Library Cataloguing in Publication Data

A catalogue record for this book is available from the British Library

ISBN 0471966932

Typeset in 10/12pt Times by Pure Tech India Ltd, Pondicherry
Printed and bound in Great Britain by Bookcraft (Bath) Ltd
This book is printed on acid-free paper responsibly manufactured from sustainable
forestation, for which at least two trees areplanted for each one used for paper production.

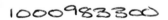

Contents

Preface

Water is one of the prime natural resources, an essential commodity for the living systems that constitute the biosphere, unique in its properties – the only substance to exist in all three phases, solid, liquid and gaseous, within the temperature range of the natural environment, vulnerable to contamination and pollution by human activities, continually renewed by the natural hydrological cycle of evaporation, vapour transportation and precipitation. Water supply and sanitation systems are primary infrastructural needs of communities worldwide, playing a key role in the promotion of public health and the elimination of disease, a fact recognised by the United Nations in designating the 1980s as the Water Supply and Sanitation Decade. Water is of course put to many other uses in addition to its drinking water and sanitation uses. High purity waters are used, for example, in the pharmaceutical and semi-conductor industries. Water is widely stored and distributed for the irrigation of crops in regions of low rainfall and is used in the electrical power industry for hydropower generation, energy storage and cooling. Thus the engineering associated with water resources management and use is multi-faceted.

This book deals with the technologies used to control and modulate water quality to meet the regulatory standards that govern its use for drinking and sanitation purposes and its related uses in industrial manufactures. It deals with the range of treatment processes used in the production of drinking and other high quality waters, in the treatment of municipal and industrial wastewaters and in the treatment and disposal of sludge residues. In general, the presentation of the subject matter proceeds sequentially from basic principles through analytical/experimental methods to the development of process design methodologies. Processes are treated as unit operations, emphasising those process fundamentals which can be applied to all process applications.

Chapter 1 presents basic information on water as a substance, with particular reference to the physical and chemical properties of relevance to process technology and design. It includes interactive computer programs which contain data bases on water properties and the solubility of gases.

Chapters 2 and 3 cover the behaviour of particulate matter in water, including particle settling behaviour, the characteristics of colloidal systems and suspensions and methods for their destabilisation. Chapters 4, 5 and 6 relate to particle removal processes, including sedimentation, flotation and filtration.

Chapter 7 reviews dissolved species in water. Chapters 8, 9 and 10 relate to dissolved species removal by adsorption, chemical precipitation and ion exchange.

Chapter 11 presents the fundamentals of biological processes and serves as a foundation for the following Chapters 12 and 13 which deal with aerobic suspended floc and biofilm processes technology and design, and for Chapter 15 which deals with the corresponding anaerobic process technologies.

Chapter 14 covers the topic of gas-liquid trasfer, the more important process applications of which are to be found in the aeration systems of activated sludge processes and in the air saturation systems associated with dissolved air flotation processes.

The characteristics, treatment and disposal of the sludge residues derived from the various processes used in water and wastewater treatment are the subject matter of Chapter 16.

Chapter 17 covers the topic of water disinfection which is of critical importance in preventing the spread of waterborne disease.

This book has been developed from lecture notes for a course of the same title presented by the author as an elective course to final year undergraduate civil engineering students and also as a core course to graduate students in water and environmental engineering at University College Dublin. Its broad coverage of process technology and design should also make it a useful reference work for engineers and applied scientists working in the general field of water and environmental engineering.

I have endeavoured to acknowledge throughout the text the numerous sources on which I have drawn in the compilation of this book. In this context, I would like to make a special reference to the seminal 2-volume publication on the same subject, authored by Fair, Geyer and Okun, published by John Wiley Inc. in 1968, which has had a formative influence on my approach to the subject. The helpful advice and comments of colleagues, including Peter O'Connor, Patrick Purcell and Michel Davitt are also gratefully acknowledged. I wish, in particular, to acknowledge the valuable contribution of Ita Casey who assisted in the preparation of the figures and John Casey who reviewed the manuscript.

tjc
Dublin
April 1996.

1

Elements of Water Science

1.1 INTRODUCTION

Water is a vital commodity with many remarkable and unique properties. In his elementary textbook on General Chemistry, Pauling (1953) states: 'Water is one of the most important of all chemical substances. It is a major constituent of living matter and of the environment in which we live. Its physical properties are strikingly different from those of other substances, in ways that determine the nature of the physical and biological world.'

Water constitutes about 70% of the human body and over 80% of most vegetables. It is the primary biological fluid, being the physico-chemical reagent that facilitates the metabolic reactions through which food is converted to energy and new cell material. It is the transport medium that removes waste metabolites from the body and it is the medium that cools the body through the thermodynamic mechanisms of transpiration, perspiration and respiration.

Water plays an essential role in photosynthesis; this is the process in which water and carbon dioxide are, by the energy of the sun, converted to carbohydrates in green plants, releasing oxygen at the same time. Photosynthesis maintains the level of oxygen in the atmosphere by replacing that consumed by all living organisms; it is also an important provider of dissolved oxygen to rivers and lakes, where it replaces that used up in the biological degradation of organic matter, thus facilitating the process of self-purification.

Because of its remarkable solvent properties and its ability to entrain and transport particulate matter, water is very susceptible to contamination. It is universally used, domestically and industrially, as a transport medium for waste, generating as a consequence large volumes of wastewater.

This textbook outlines the science and technology of the individual processes that are used for the treatment of water to upgrade its quality to meet specific water quality standards. This 'unit operations' approach emphasizes the common process fundamentals whether used in drinking water production or wastewater treatment systems.

1.2   WATER CHEMISTRY

1.2.1   Structure and composition of water

The experiments of Cavendish and Lavoisier in the 1780s established that water is composed of hydrogen and oxygen in the ratio of two volumes of hydrogen to one volume of oxygen, i.e. the molecular formula for water is H_2O with a molecular weight of 18. Isotopic analysis (Eisenberg and Kaufman, 1969) has indicated, however, that many natural waters contain a very small rare isotope fraction which generally does not exceed 0.3%. (There are at present three known isotopes of hydrogen, ^1H, ^2H (deuterium), ^3H (tritium), and six of oxygen, ^{14}O, ^{15}O, ^{16}O, ^{17}O, ^{18}O and ^{19}O. Tritium is radioactive with a half-life of 12.5 years; the isotopes ^{14}O, ^{15}O and ^{19}O are also radioactive, but are short-lived and do not occur significantly in natural water.)

The water molecule may be described as a 'bent' molecule with an atomic configuration, as shown schematically in Figure 1.1. In simple terms, its electronic configuration may be described as consisting of four electron orbitals in which four pairs of the outer electrons (six from oxygen and one each from the two hydrogens) are likely to be found. These orbitals are arranged approximately along the directions joining the oxygen atom to the corners of a tetrahedron. Of the four electron orbitals, two are used for the covalent O–H bonds and the remaining two are free orbitals for the remaining so-called lone electron pairs. This electronic configuration confers on the water molecule the important characteristic of charge separation i.e. the resultant centres of positive and negative charge do not coincide. The electrically neutral water molecule can thus be considered as an electric dipole, which would be subject to an orientating moment in an electric field. The moment of the dipole is defined as the product of the electric charge by the distance between the charge centres (units esu.cm). In the gas phase the dipole moment of water is 1.87×10^{-18} esu.cm and, owing to molecular association, it becomes somewhat greater in the liquid phase.

Molecular association, or 'hydrogen bonding', which occurs characteristically in ice crystals and is also thought to exist in liquid water, is a

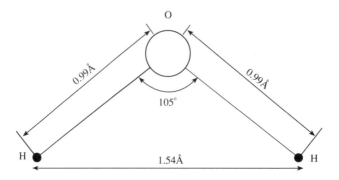

Figure 1.1
Schematic configuration of the water molecule

further important consequence of the charge separation on the water molecule. Because of this separation the oxygen atom of the water molecule exerts electrostatic attractive forces, in directions corresponding to the orbitals of its lone electron pairs, on the oppositely charged hydrogen atoms of other water molecules. This electrostatic bonding is referred to as hydrogen bonding and, because of it, the structure of ice is arranged in an ordered lattice framework, in which each oxygen atom is tetrahedrally surrounded by four other oxygen atoms. In between any two oxygen atoms is a hydrogen atom which provides the hydrogen bonding. Such a structure of associated water molecules contains interstitial regions which are larger than the dimensions of the water molecule and hence permits the coexistence of free unassociated water molecules.

In liquid water the ordered molecular structure of ice is considered to be partly retained (Pauling, 1960). Its structure may be visualized as consisting of clusters of associated molecules, hydrogen bonded in a tetrahedral framework, plus free unassociated molecules, with continuous interchange between the two. The hydrogen bond is weak relative to the co-valent O–H bond. It has a bond energy of 4.5 kcal mole^{-1} as compared with a value of 110 kcal mole^{-1} for the O–H bond. For this reason, the hydrogen bond is easily broken.

The water molecule dissociates to form hydrogen (H^+) and hydroxyl (OH^-) ions. As will be discussed later, these ions play a major role in aqueous chemical reactions. In combination with the polar nature of the water molecule they are responsible for the remarkable solvent properties of water.

1.2.2 Dissolved substances in water

Substances dissolved in water (solutes) are dispersed in either molecular or ionic form. The concentration of such substances can be expressed in a variety of units (see Table 1.1)

Table 1.1

Basis	Unit
Weight/volume (w/v)	mg l^{-1}, kg m^{-3}
Weight/weight (w/w)	mg kg^{-1}, ppm
Molarity	moles l^{-1}
Molality	moles kg^{-1}
Normality	equivalents l^{-1}

In many water and wastewater treatment situations, the solution density is effectively the same as pure water (in practical engineering computations taken to be 1000 kg m^{-3}; refer to Table 1.3 for precise values). Hence, the numerical value of the solute concentration, whether expressed as mg l^{-1}, ppm or mg kg^{-1}, is effectively the same. By common usage and method of measurement, the w/v form as mg l^{-1} is preferred for waters and wastewaters, while the w/w form is commonly used in practice in relation to concentrated suspensions such as sludges.

The designation of solute concentration in molar units is particularly appropriate for stoichiometric computations relating to aqueous chemical reactions. A one molar solution of a substance contains one mole of the substance per litre of solution, a mole being the mass of a substance equal to its molecular weight in grams, e.g. a mole of oxygen has a mass of 32 g.

The designation of solute concentration in normality terms is convenient for chemical reactions relating to (1) ion charge, (2) acid-base reactions that involve hydrogen or hydroxyl ion transfer, and (3) oxidation-reduction reactions which involve electron transfer. A one normal solution of a substance is a solution that contains one 'equivalent weight' of substance per litre of solution. The equivalent weight of a substance is its molecular weight divided by a factor, the value of which is determined by whichever one of the above three chemical environments exists, as illustrated by the following examples:

(1) Ion charge. Iron precipitates from an aqueous solution through the reaction

$$Fe^{3+} + PO_4^{3-} \rightarrow FePO_4 \, (s) \tag{1.1}$$

The ferric iron ion has three charges; hence the equivalent weight of iron   is its molecular weight/3 = 55.8/3 = 18.6 g. If the iron concentration in solution is 50 mg l^{-1}, then the normality is calculated as follows:

$$\text{normality} = \frac{50 \times 10^{-3}}{18.6} = 2.69 \times 10^{-3} \text{ eq} \, l^{-1} = 2.69 \text{ meq} \, l^{-1}$$

(2) Acid-base reactions. Phosphoric acid dissociates in solution according to the reaction

$$H_3PO_4 \Leftrightarrow 2H^+ + HPO_4^{2-} \tag{1.2}$$

Two hydrogen ions are released and hence the equivalent weight of phosphoric acid is its molecular weight/2 = 98/2 = 49 g. If the concentration of phosphoric acid in solution is 50 mg/l, then the normality of the solution is calculated as follows:

$$\text{normality} = \frac{50 \times 10^{-3}}{49} = 1.02 \times 10^{-3} \text{ eq} \, l^{-1} = 1.02 \text{ meq} \, l^{-1}$$

(3) Oxidation-reduction reactions. Ferrous iron is oxidized to ferric iron according to the reaction

$$Fe^{2+} \rightarrow Fe^{3+} + e^- \tag{1.3}$$

One electron is released and hence the equivalent weight of iron is equal to its molecular weight. If the concentration of iron is 50 mg l^{-1}, then the normality of the solution is calculated as follows:

$$\text{normality} = \frac{50 \times 10^{-3}}{55.8} = 0.9 \times 10^{-3} \text{ eq} \, l^{-1} = 0.9 \text{ meq} \, l^{-1}$$

1.2.3   Organic matter in water

The concentration of organic matter in water is commonly expressed in terms of the oxygen or equivalent oxygen consumed in its oxidation to carbon dioxide, i.e. in terms of oxygen demand.

The biochemical oxygen demand, or BOD_5, is defined as the oxygen uptake in microbial respiration over a 5-day period during which the water sample is incubated out of contact with light, in a constant temperature environment of 25°C. Essentially, BOD is an indirect measure of biodegradable organic matter.

The chemical oxygen demand, or COD, is the oxygen-equivalent of the dichromate consumed in the chemical oxidation of organic matter to carbon dioxide. In the dichromate oxidation-reduction reaction, hexavalent chromium is reduced to trivalent chromium with the loss of three electrons. The equivalent amount of oxygen is computed on the basis that each mole of oxygen accepts four moles of electrons in its reduction to carbon dioxide. The COD is an indirect measure of the concentration of chemically oxidizable organic matter present in water.

Other measures of organic matter in water that are commonly used in water and wastewater treatment processes discussion are total organic carbon, or TOC, and dissolved organic carbon, or DOC. These are derived from direct instrumental measurements of the amount of carbon dioxide produced through the oxidation of the organic carbon present in a water sample.

1.2.4   Chemical reactions

Many of the water and wastewater treatment processes discussed later in this work, are based on the use of chemical reactions to effect a particular change in water quality. Examples include the chemical coagulation process, which is widely used in the production of drinking water to remove colour and turbidity-causing colloidal matter from water; the softening of hard waters to remove calcium and magnesium by chemical precipitation using lime; the removal of heavy metals from wastewaters by chemical precipitation, and many others. In general, the two factors that are of special importance for the process designer are the rate at which reactions take place and the residuals that remain in solution/suspension in the treated water or wastewater. These factors are largely determined by the principles of reaction kinetics and system equilibrium.

Process kinetics (Snoeyink and Jenkins, 1980) is concerned with the rate at which reactions take place and the environmental factors that influence the reaction rates. In general, the rate of chemical reaction is related to the concentrations of reaction species, which can be stated in mathematical terms as follows:

for the irreversible reaction : $\qquad$ A + 2B $\qquad \rightarrow \qquad$ C + 2D

$\qquad\qquad\qquad\qquad\qquad\qquad$ reactants $\qquad\qquad$ products

rate law :
$$\frac{d[A]}{dt} = -k[A]^a[B]^b[C]^c[D]^d \qquad (1.4)$$

where $[\cdot]$ indicates the molar concentration, k is the reaction rate constant, and a, b, c and d are empirical constants.

The value of the rate constant, k is found to increase with increasing temperature according to the semi-empirical Arrhenius relationship:

$$k = Ae^{-(E_a/RT)}$$

where A is a constant for a particular reaction; E_a is the activation energy, which may also be regarded as a constant for a particular reaction; R is the ideal gas constant and T is the absolute temperature (K).

The rate at which some reactions proceed may be greatly accelerated by the presence of a catalyst, which effectively facilitates the reaction without direct participation, i.e the concentration of the catalyst is not changed by the reaction. Hydrogen ions (H^+) and hydroxyl ions (OH^-) are common catalysts in aquatic systems, as reflected in the extent to which the rates of many reactions are influenced by the reaction pH. As illustrated in Chapter 17, cobalt ion is used as a catalyst to accelerate the rate of oxidation of the sulphite ion (SO_3^{2-}) to sulphate (SO_4^{2-}) by oxygen in the chemical de-oxygenation of water in carrying out oxygenation capacity tests on aeration systems.

As already noted, the residuals that remain in solution following a chemical reaction process are influenced both by the reaction rate and the solubility of the species involved in the process. The solubility of a substance in water can be defined as its equilibrium concentration in an aqueous solution containing an excess of the substance as a precipitate. The equilibrium state is a dynamic condition in which the forward and reverse rates in a reversible reaction are in balance. Consider the reversible reaction:

$$aA + bB \Leftrightarrow cC + dD$$

rate of the forward reaction: $v_f = k_1[A]^a[B]^b$ $\qquad\qquad\qquad$ (1.6)

rate of reverse reaction: $v_r = k_2[C]^c[D]^d$ $\qquad\qquad\qquad$ (1.7)

At equilibrium these reactions are in balance, i.e. $v_f = v_r$ and hence

$$\frac{[C]^c[D]^d}{[A]^a[B]^b} = \frac{k_1}{k_2} = K \tag{1.8}$$

where K is the equilibrium constant.

Where the reaction is a precipitation reaction, the equilibrium constant is generally called the 'solubility product'. Take, for example, the aqueous equilibrium relationship of the precipitate AB and its dissolved ions:

reaction: $\qquad\qquad A_aB_{b(s)} \Leftrightarrow aA^{b+} + bB^{a-}$

at equilibrium: $\qquad\dfrac{[A^{b+}]^a[B^{a-}]^b}{[A_aB_{b(s)}]} = K \tag{1.9}$

By convention the activity of the precipitate is taken to be unity and K is denoted as K_{so} the so-called solubility product. Hence

$$K_{so} = [A^{b+}]^a[B^{a-}]^b \tag{1.10}$$

Solubility product values for a range of precipitates of importance in water treatment technology are given in Table 3.2 in Chapter 3.

One of the key equilibrium relationships in water is its dissociation to form hydrogen and hydroxyl ions:

$$H^+ + OH^- \Leftrightarrow H_2O$$

This reaction proceeds very rapidly so that equilibrium conditions are quickly attained in accordance with the equilibrium condition

$$[H^+][OH^-] = 10^{-14} \tag{1.11}$$

The hydrogen ion concentration is conventionally expressed in terms of the negative logarithm of its molar concentration or pH. Equilibrium constants are also commonly expressed in pK units.

Another important basic equilibrium relationship that universally applies in aqueous solutions is that of electroneutrality, or charge balance. For example, if the ionic species present are H^+, OH^-, M^{x+} and A^{y-}, then the corresponding charge balance condition is

$$[H^+] + x[M^{x+}] = [OH^-] + y[A^{y-}] \tag{1.12}$$

Molar concentration has been the unit of concentration expression used throughout the foregoing discussion of elementary aspects of chemical kinetics and chemical equilibrium. It should be noted, however, that where reactions involve concentrated solutions the molar concentrations have to be modified by activity coefficients to take into account the effects of high concentration on reaction rate and reaction equilibrium. In water treatment systems, however, we are almost invariably dealing with very dilute aqueous solutions.

1.2.5 The carbonate system

The carbonate system plays a key role in the chemical stability of waters, influencing pH, buffer capacity and corrosivity. The carbonate species in natural and treated waters include:

free carbon dioxide	$CO_{2(aq)}$
carbonic acid	H_2CO_3
bicarbonate ion	HCO_3^-
carbonate ion	CO_3^{2-}

Carbon dioxide is taken into solution from the atmosphere; it is also produced in solution by respiring organisms, while it is consumed by photosynthetic organisms. Its equilibrium, or saturation concentration, in water in contact with atmospheric air is given by the equilibrium relation:

$$[CO_{2(aq)}] = K_h P_{CO_2}$$

where $[CO_{2(aq)}]$ is the molar concentration of free carbon dioxide in solution, K_h is the Henry's law constant (see Table 1.2), and P_{CO_2} is the partial pressure (atm) of carbon dioxide in the atmosphere ($P_{CO_2} = 10^{-3.5}$ atm in normal atmospheric air).

The reactions between the carbonate species in water and the related equilibrium constants are as follows:

	Reaction	Equilibrium constant
(a)	$CO_{2(aq)} + H_2O \Leftrightarrow H_2CO_3$	K_a
(b)	$H_2CO_3 \Leftrightarrow H^+ + HCO_3^-$	K_b
(c)	$HCO_3^- \Leftrightarrow H^+ + CO_3^{2-}$	K_2

Since only about 0.16% of the aqueous CO_2 forms carbonic acid, it is convenient for computation purposes to combine $CO_{2(aq)}$ and H_2CO_3, denoting their combination as $H_2CO_3^*$:

$$[CO_{2(aq)}] + [H_2CO_3] = [H_2CO_3^*]$$

Reaction (b) is therefore more usually expressed in the form:

(d)	$H_2CO_3^* \Leftrightarrow H^+ + HCO_3^-$	K_1

Carbonate is removed from solution through precipitation, most commonly as calcium carbonate:

$CaCO_{3(s)} \Leftrightarrow Ca^{2+} + CO_3^{2-}$	K_{sp}

where the solubility product $K_{sp} = [Ca^{2+}][CO_3^{2-}]$.

The values of the more commonly used aqueous carbonate system constants in the temperature range 0–60°C are given in Table 1.2 (it should be noted that $K_1 \cong K_a \times K_b$).

Table 1.2 Carbonate system equilibrium constants in pK units

Equilibrium constant	Temperature (°C)						
	5	10	15	20	25	40	60
K_h	1.20	1.27	1.34	1.41	1.47	1.64	1.80
K_1	6.52	6.46	6.42	6.38	6.35	6.30	6.30
K_2	10.56	10.49	10.43	10.38	10.33	10.22	10.14
K_{sp}	8.09	8.15	8.22	8.28	8.34	8.51	8.74

1.3 WATER PHYSICS

The physical properties of water, particularly those that influence flow, mixing and turbulence, have a major bearing on the performance of water and wastewater treatment processes. The properties that influence the flow behaviour of all fluids include density, viscosity and surface tension.

1.3.1 Viscosity

The viscosity of a fluid is the property that defines its resistance to flow. It is quantified in terms of a coefficient of dynamic viscosity μ, which is defined by the relationship:

$$\tau = \mu \frac{dv}{dy} \tag{1.13}$$

where τ is the shear stress associated with the velocity gradient dv/dy in a flow environment in which turbulence is suppressed. The units of dynamic viscosity are $N\,s\,m^{-2}$. The ratio of dynamic viscosity to density (μ/ρ) commonly appears in flow computations; it is known as the kinematic viscosity, has units of $m^2\,s^{-1}$ and is generally designated by the symbol ν.

The magnitude of the coefficient of dynamic viscosity μ for liquids decreases with an increase in temperature. Dynamic viscosity values for water in the temperature range 0–100°C are given in Table 1.3.

The fluids that exhibit the foregoing flow behaviour are known as Newtonian fluids. They include waters, wastewaters and gases. The linear correlation of shear stress and velocity gradient, characteristic of Newtonian fluids, prevails only in the absence of turbulence in the flow field. This type of flow is described as *laminar* flow and, for Newtonian fluids, is confined to situations where random bulk fluid movement is suppressed as, e.g. flow in small bore pipes or through porous media or very close to solid boundaries. Where turbulence exists in the flow, however, the shear resistance is greatly increased and the associated shear stress can, for convenience, be correlated to the velocity gradient by an expression of the same form as that used to define dynamic viscosity:

$$\tau = \varepsilon \frac{\mathrm{d}v}{\mathrm{d}y} \qquad (1.14)$$

where ε is the coefficient of *eddy viscosity* and is a characteristic of the flow, as distinct from μ which is a property of the fluid. The coefficient of eddy viscosity may be regarded as a coefficient of momentum transfer along the velocity gradient; its magnitude is dependent on the velocity gradient, shear stress, and other factors and is invariably much greater than the dynamic viscosity, μ.

Unlike water and gases, sludges exhibit non-Newtonian behaviour, particularly at high concentration. Non-Newtonian flow behaviour is characterized by a non-linear relation of shear stress and velocity gradient or rate of shear strain, and, in some fluids, by the existence of a yield stress which must be exceeded for flow to take place (Casey, 1992).

1.3.2 Surface tension

The interfacial liquid at the boundary between a liquid and a gas behaves rather like a membrane which possesses tensile strength. This membrane-like behaviour can be quantified as a strain energy per unit area, i.e. $N\ m\ m^{-2}$ or force per unit length ($N\ m^{-1}$), denoted by the symbol σ.

Surface tension causes the capillary rise of water above the phreatic surface in fine-grained saturated soils and porous construction materials. The surface tension influence is generally very small in most fluid flow situations encountered in water and wastewater treatment systems.

The surface tension of water decreases with an increase in temperature. Surface tension values for water in the temperature range 0–100°C are given in Table 1.3.

1.3.3 Vapour pressure

When evaporation takes place from the surface of a liquid within an enclosed space or vessel, the partial pressure created by the vapour molecules is called vapour pressure. A liquid may, at any temperature, be considered to be in equilibrium with its own vapour when the rate of molecular transport through the separating gas–liquid interface is the same in both directions. The absolute pressure corresponding to this concentration of gas molecules is defined as the saturation pressure of the the liquid. The saturation vapour pressure of every liquid increases with an increase in temperature. The temperature at which it reaches 1 atm absolute is the boiling point, which for water is 100°C.

Saturation pressure values for water in the temperature range 0–100°C are given in Table 1.3.

Table 1.3 Physical properties of water

Temperature (°C)	Density (kg m^{-3})	Saturation vapour pressure (N m^{-2} × 10^{-3})	Dynamic viscosity (N s m^{-2} × 10^3)	Surface tension (N m^{-1} × 10^3)
0	999.87	0.6107	1.787	75.64
5	999.99	0.8721	1.519	74.92
10	999.73	1.2277	1.307	74.22
15	999.13	1.7049	1.139	73.49
20	998.23	2.3378	1.002	72.75
25	997.07	3.1676	0.890	71.97
30	995.68	4.2433	0.798	71.18
35	994.06	5.6237	0.719	70.37
40	992.25	7.3774	0.653	69.56
45	990.24	9.5848	0.596	68.74
50	988.07	12.3380	0.547	67.91
55	985.73	15.7450	0.504	67.05
60	983.24	19.9240	0.467	66.18
65	980.59	25.0130	0.434	65.29
70	977.81	31.1660	0.404	64.40
75	974.89	38.5530	0.378	63.50
80	971.83	47.3640	0.355	62.60
85	968.65	57.8080	0.334	61.68
90	965.34	70.1120	0.315	60.76
95	961.92	84.5280	0.298	59.84
100	958.38	101.3250	0.282	58.90

Source: *CRC Handbook of Chemistry and Physics*, 67th edn (1987).

1.3.4 Density

Data on pure water density confirm that water has an open-type structure. Bernal and Fowler (1933) estimated that water with a closely packed molecular structure would have a density some 84% greater than the observed value.

Unlike other liquids, water at atmospheric pressure has a negative thermal expansion between 0°C and 4°C. This anomaly is considered to be related to the competing effects of (a) the breakdown of hydrogen bonding, resulting in a volume reduction, and (b) the increase in the amplitude of molecular vibrations, causing an increase in volume. Between 0°C and 4°C the volume decrease due to structure breakdown is greater than the volume increase due to increased molecular activity, resulting in a net reduction in volume and corresponding increase in density. Above 4°C the vibrational effect is considered to outweigh the structural effect and water density increases with increasing temperature.

Values of pure water density at atmospheric pressure in the temperature range 0–100°C are given in Table 1.3.

While water may be regarded as incompressible in relation to computations involving its density or weight, it is of course an elastic material with a coefficient of compressibility (or bulk modulus) of about 21.1×10^8 N m^{-2} at 10°C. This value increases marginally with increase in temperature and pressure. This property is of importance in relation to the transmission of pressure waves through water as, for example, in the so-called 'waterhammer' effect caused by rapid deceleration of flow in closed conduits.

The densities of natural waters and wastewaters are greater than pure water at the same temperature. In general, the difference is only of practical significance where the concentration of dissolved or suspended material is very high.

The extent by which the densities of aqueous solutions exceed that of pure water is a complex function of salt concentration, temperature and pressure. For practical process engineering purposes, the increase in density over pure water (see Table 1.3) may be approximated as follows:

$$\Delta\rho = 0.75S \qquad (1.15)$$

where $\Delta\rho$ is the density increase (kg m^{-3}) over pure water at the same temperature and S is the total salt concentration (kg m^{-3}). Thus the density of brackish water at 10°C, having a salt concentration of 4 kg m^{-3}, is $999.73 + 4 \times 0.75 = 1002.73$ kg m^{-3}

The density of an aqueous suspension is a function both of the gravimetric concentration of suspended solids and the specific gravity of the solid matter contained in the suspended particles. Consider a suspension of a mass M containing $p\%$ (w/w) solid matter of dry density ρ_s: suspension density:

$$\rho_{susp} = \frac{mass}{vol} = \frac{M}{\left(\dfrac{p}{100}\dfrac{M}{\rho_s}\right) + \left(\dfrac{100-p}{100}\dfrac{M}{\rho}\right)}$$

Hence

$$\rho_{susp} = \frac{\rho}{1 - \dfrac{p}{100}\left(1 - \dfrac{1}{S_g}\right)}$$

where ρ is the density of water and S_g is the specific gravity (ρ_s/ρ) of the dried suspended solids.

1.3.5 Diffusivity

The random movement (Brownian motion) of ions, molecules and very small particles (colloids) in water is known as diffusion. The mean velocity of such species conforms approximately to the basic equation of the simple kinetic theory (Camp and Meserve, 1974):

$$\tfrac{1}{3}Nmu^2 = R_u T \qquad (1.17)$$

where N is Avogadro's number (the number of molecules per mole $= 6.06 \times 10^{23}$), m is the mass of each particle (kg), u is the mean velocity of the particles (m s^{-1}), and R_u is the universal gas constant, which has the value 8.3144 J mole^{-1} K^{-1} for a perfect gas, T is the absolute temperature (K). It follows from equation (1.17) that the kinetic energy of solutes and small particles is independent of their size and that their mean velocity is inversely related to the square root of their mass, i.e. the smaller the particle mass the greater is its mean velocity.

The rate of diffusion is defined by the following equation, which is generally known as Fick's law (Crank, 1956):

$$\frac{\partial m}{\partial t} = -D\frac{\partial c}{\partial x}$$

where $\partial m/\partial t$ is the rate of mass transfer per unit area in the x-direction, $\partial c/\partial x$ is the concentration gradient in the x-direction of the diffusing species, and D is the diffusion coefficient or diffusivity, the magnitude of which defines the mass diffusion rate. The units of D are length^2/time, usually expressed as cm^2 s^{-1}. Some aqueous diffusion coefficient values are presented in Table 1.4.

Table 1.4 Diffusion coefficients in pure water

Solute	Molal concentration	Temperature (°C)	Diffusivity, D (cm^2 s^{-1})
Dissolved oxygen (O$_2$)	—	20	2.03×10^{-5}
Dissolved nitrogen	—	20	1.88×10^{-5}
Hydrochloric acid (HCl)	0.1	19	2.56×10^{-5}
Sodium chloride (NaCl)	0.1	15	1.09×10^{-5}
Calcium chloride	2.0	10	0.79×10^{-5}
Chlorine (Cl$_2$)	0.1	12	1.41×10^{-5}
Sulphuric acid (H$_2$SO$_4$)	1.0	12	1.30×10^{-5}
Magnesium sulphate (MgSO$_4$)	1.0	7	0.35×10^{-5}
Ammonia (NH$_3$, NH$_4^+$)	1.0	15	1.78×10^{-5}
Acetic acid (CH$_3$COOH)	0.2	13.5	0.89×10^{-5}
Methyl alcohol (CH$_3$OH)	—	18	1.37×10^{-5}
Glycerol (C$_3$H$_5$OH$_3$)	—	20	0.83×10^{-5}
Glucose (C$_6$H$_{12}$O$_6$)	—	18	0.57×10^{-5}
Urea (NH$_2$CONH$_2$)	—	20	1.18×10^{-5}
Various dyes	—	18	$(0.17–0.58) \times 10^{-5}$

Source: Camp and Meserve (1974).

It is important to note that diffusion is the dominant transport mechanism in aqueous systems only where turbulent bulk water movement is suppressed. This is rarely the case in water and wastewater treatment systems. Thus, for example, while the rate of oxygen transfer from air into surface waters is largely controlled by a diffusion process in the interfacial water film, its transport into the water body is much more rapid, so much so that the oxygen concentration variation in the bulk water is generally negligible.

1.3.6 Water-dispersed particles

Water-dispersed species range in size from *true solutes*, i.e. ions and molecules to suspended particles visible to the naked eye. Figure 1.2 categorizes the more significant water- dispersed species by size range.

Ions and inorganic molecules may be regarded as true solutes, ranging in size up to about 10 Å or 10^{-3} µm. Low molecular weight organic molecules such as glucose (MW = 180) also belong in this category. High molecular weight organics such as proteins and polymers are an order of magnitude larger and fall within the size range 10^{-3}–10^{-2} µm.

Dispersed particles that are smaller than about 1 µm and larger than true solutes are designated as *colloids* and exhibit characteristic colloi-

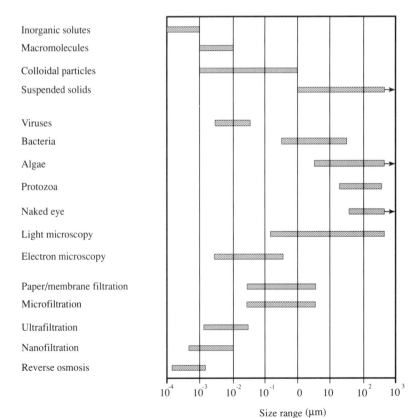

Figure 1.2
Characteristics of water-dispersed species

dal properties. As indicated in Figure 1.2, they can be observed only by means of an ultra microscope or an electron microscope. Although colloids are too small to be distinguished by the naked eye, when viewed at right angles to a light beam, however, colloidal outlines can be distinguished (Tyndall effect). As briefly discussed in section 1.3.5, colloidal particles are subject to random Brownian motion, having a mean velocity that conforms approximately to equation (1.17). It follows from this equation that the average kinetic energy of colloidal particles ($\frac{1}{2}mu^2$) is independent of their size. Thus, the mean velocity of the smallest colloidal particle (10^{-3} µm) is calculated from equation (1.15) to be about 100 m s^{-1}, while the velocity of the largest colloidal particle (1 µm) is calculated to be only about 0.004 m s^{-1}. Colloidal dispersions have a high degree of stability and are not amenable to removal by sedimentation or conventional filtration processes. The nature of colloidal stability and the processes used to bring about destabilisation are discussed in Chapter 3. The uniform kinetic energy of colloidal particles, resulting in the very small colloidal particles having very high mean velocities, is a critical factor in the destabilization process.

There is no direct method of measuring the concentration of colloidal particles in water. Turbidity measurement, however, provides a useful index of colloid concentration. Turbidity or cloudiness in water results from the light-scattering effect of very small particles, the outline of which is not visible to the naked eye. It is measured (APHA Standard Methods, 1992) by reference to a standard aqueous turbidity range produced by the dispersion of formazin in water of zero turbidity. Turbidity is normally expressed in nephelometric turbidity units (NTU).

The dispersed species, known as *total suspended solids* (TSS), may be loosely regarded as those solids greater in size than colloids. The TSS concentration is measured by filtration through glass-fibre filter paper having a pore size in the range 1–1.6 µm, the retained residue being dried to constant weight at a temperature of 103–105°C.

The 'fixed' suspended solids concentration is the residue remaining after the dried TSS residue is ignited to constant weight at a temperature of (550 ± 50°C. The loss on ignition at this temperature represents the volatile suspended solids.

Settleable solids constitute a subset of TSS, being that fraction which is separated to form a visibly defined settled solids residue after settling for 1 h. The settleable solids concentration may be expressed in volumetric (ml l^{-1}) or gravimetric (mg l^{-1}) units. The volumetric measurement of settleable solids is made using an Imhoff cone in which a 1 l sample is allowed to settle under quiescent conditions for a period of 1 h; the settleable solids concentration in volumetric terms is represented by the volume occupied by the settled solids at the base of the cone. The gravimetric concentration is determined as the difference between the measured TSS of a test sample and the TSS remaining in the supernatant after a 1 h period of quiescent settling.

Living organisms also constitute an important category of dispersed matter in waters and wastewaters. While the larger algae and protozoa are of a size that can be observed by the naked eye (see Figure 1.2), most microorganisms, with the exception of viruses, can be observed using a light microscope and can be separated from water by membrane filtration techniques. Viruses, however, are an order of magnitude smaller than the smallest bacteria. They can be observed only under the magnification of the electron microscope and require filters in the ultrafiltration range for their separation from waters and wastewaters.

1.4 GAS–WATER INTERACTIONS

A number of the unit treatment processes used in water and wastewater treatment systems involve a gas phase. These processes include aeration, dissolved air flotation, chlorination, anaerobic digestion and decarbonation. The gases of relevance include oxygen, ozone, nitrogen, carbon dioxide, air, methane, hydrogen sulphide, ammonia, chlorine and sulphur dioxide.

1.4.1 Gas properties

The *thermodynamic* properties of a gas govern the interrelation of pressure, volume and temperature, which for most gases is defined by the equation of state for the so-called perfect gas, usually written in its general form as follows:

$$PV = mR_u\Theta \qquad (1.19)$$

where P is the absolute pressure (N m^{-2}), V is the gas volume (m^3), m is the mass of gas (mole), R_u is the universal gas constant (J mole^{-1} K^{-1}), and Θ is the absolute temperature (K). The perfect gas has an R_u-value of 8.3144 J mole^{-1} K^{-1}.

Changing from mole to kg, equation (1.19) may be written for individual gases in the form:

$$\frac{P}{\rho} = R\Theta \qquad (1.20)$$

where ρ is the gas density (kg m^{-3}) and R is the specific gas constant (J kg^{-1} K^{-1}), related to R_u as follows:

$$R = \frac{1000\, R_u}{w} \qquad (1.21)$$

where w is the molecular weight.

The constant R can be shown to be the difference between the specific heat capacity of a gas at constant pressure (C_p) and its specific heat capacity at constant volume (C_v). Values for these thermodynamic properties are given in Table 1.5.

The relationships embodied in equations (1.19) and (1.20) may also be expressed in the forms:

$$PV^\gamma = \text{constant} \tag{1.22}$$

or

$$\frac{P}{\rho^\gamma} = \text{constant} \tag{1.23}$$

Table 1.5 Thermodynamic properties of gases

Gas	C_p (J kg^{-1} K^{-1})	C_p/C_v	R (J kg^{-1} K^{-1})
Air	1005.0	1.40	287.1
Oxygen	920.0	1.40	262.9
Nitrogen	1040.0	1.40	297.1
Methane	2260.0	1.31	534.8
Carbon dioxide	867.0	1.30	202.2

Values relate to 25°C and 1 atm; K = °C + 273.15.
Source: *CRC Handbook of Tables for Applied Engineering Science*, 2nd edn (1976).

where V is the gas volume (m^3) and γ is the so-called polytropic exponent. The value of γ depends on the process by which the gas undergoes volume change. For adiabatic processes (zero internal energy loss), is equal to the specific heat ratio C_p/C_v, whereas for isothermal processes (zero temperature change), γ is equal to unity. Thus, in real situations, the value of γ lies within the range 1.0 to C_p/C_v.

Gases are relatively highly compressible, their compressibility depending on temperature and pressure. The coefficient of compressibility K for a gas is given by the relation:

$$K = P\gamma \tag{1.24}$$

Thus, the coefficient of compressibility, which is a measure of the stiffness of a gas, increases with pressure, which means that gases become less compressible with increasing pressure.

The *dynamic viscosity* of gases increases with temperature according to the following empirical correlation (Maitland and Smith, 1972):

$$\ln\left(\frac{\mu}{S}\right) = A \ln \Theta + \frac{B}{\Theta} + \frac{C}{\Theta^2} + D \tag{1.25}$$

where μ is the dynamic viscosity (N s m^{-2}) at temperature Θ (K); S is the dynamic viscosity (N s m^{-2}) at a standard temperature of 293.2 K; and A, B, C and D are coefficients determined from a least-squares regression analysis. Recommended values for these coefficients are presented in Table 1.6.

Table 1.6 Dynamic viscosity coefficients for gases

Gas	A	B	C	D	S (N s m$^{-2} \times 10^7$)
Air	0.63404	−45.6380	380.87	−3.4500	182
Oxygen	0.52662	−97.5893	2650.70	−2.6892	203.2
Nitrogen	0.60097	−57.005	1029.1	−3.2322	175.7
Methane	0.54188	−127.5700	4700.80	−2.6952	109.3
Carbon dioxide	0.44037	−288.4000	19312.00	−1.7418	146.7

Source: Maitland and Smith (1972).

1.4.2 Gas solubility in water

With respect to their solubility, the gases of interest in water and waste-water engineering can be split into two categories: (a) the poorly soluble gases which include oxygen (O_2), ozone (O_3), nitrogen (N_2) and methane (CH_4), and (b) gases of medium to high solubility including carbon dioxide (CO_2), ammonia (NH_3), hydrogen sulphide (H_2S), sulphur dioxide, chlorine (Cl_2) and chlorine dioxide (ClO_2).

The poorly soluble gases, namely oxygen, ozone, methane and nitrogen, conform to Henry's law which states that the solubility of a gas in a liquid with which it is in contact is proportional to the partial pressure of the gas in the gas phase:

$$c_s = H_m p \tag{1.26}$$

where c_s is the equilibrium or saturation concentration (moles l^{-1}) of the gas in solution, p is the partial pressure (atm) of the gas in the gas phase in contact with the liquid and H_m is the Henry law constant. Thus, H_m represents the gas solubility at a partial pressure of 1 atm. It is more convenient for general engineering use, however, to express gas solubility in w/v or w/w units at a gas presure of 1 atm, inclusive of the aqueous vapour pressure, as presented in Tables 1.7–1.9, inclusive. Data for the

Table 1.7 Solubilities of low-solubility gases in water in mg l^{-1} at a pressure of 1 atm, inclusive of the aqueous vapour pressure

Gas	Temperature (°C)										
	0	5	10	15	20	25	30	35	40	45	50
Oxygen (O_2)	69.9	60.9	53.8	48.1	43.4	39.4	36.1	33.2	30.8	28.5	26.5
Ozone (O_3)		41.4	35.8	29.8	24.1	18.9	13.7	9.6	6.6	4.3	2.7	1.6
Methane (CH_4)	39.8	34.4	30.0	26.4	23.7	21.5	19.9	18.3	17.1	16.1	15.4
Nitrogen (N_2)*	29.4	26.0	23.1	20.8	19.0	17.5	16.2	15.0	13.9	13.0	12.2

*Atmospheric nitrogen: 98.815% by vol. N2 + 1.185% by vol. A
Source: CRC Handbook of Chemistry and Physics (1961, 1992); Horvath (1975).

low-solubility gases are presented in Table 1.7 in units of mg l^{-1}, while data for the higher solubility gases are presented in Table 1.8 as % by weight or g per 100 g. The solubility values for these latter gases, which react with water, include all chemical species of the gas in solution. Solubility data for the atmospheric gases at an air pressure of 1 atm, inclusive of the aqueous vapour pressure, are presented in Table 1.9.

The kinetic aspects of gas transfer to/from water are discussed in detail in Chapter 14.

Table 1.8 Solubilities of high-solubility gases in water as % by wt. at a pressure of 1 atm, inclusive of the aqueous vapour pressure

Gas	Temperature (°C)						
	0	5	10	15	20	25	30
Carbon dioxide (CO_2)	0.3346	0.2774	0.2318	0.1970	0.1688	0.1449	0.1257
Chlorine (Cl_2)	—	1.160	0.990	0.850	0.730	0.640	0.567
Chlorine dioxide (CLO_2)	19.08	15.38	12.46	10.12	8.25	6.74	5.52
Ammonia (NH_3)	98.0	81.0	69.5	60.0	51.7	45.2	40.4
Hydrogen sulphide (H_2S)	0.7066	0.6001	0.5112	0.4411	0.3846	0.3375	0.2983
Sulphur dioxide (SO_2)	22.83	19.31	16.21	13.54	11.28	9.41	7.80

Source: CRC Handbook of Chemistry and Physics (1961, 1992); Horvath (1975).

Table 1.9 Solubility of atmospheric gases in water (mg l^{-1}) at an air pressure of 1 atm, inclusive of the aqueous vapour pressure

Gas	Temperature (°C)										
	0	5	10	15	20	25	30	35	40	45	50
O_2	14.64	12.76	11.27	10.08	9.08	8.25	7.57	6.95	6.45	5.97	5.61
N_2**	23.25	20.66	18.52	16.76	15.30	14.08	13.05	12.16	11.38	10.67	10.01
CO_2	1.00	0.83	0.70	0.59	0.51	0.43	0.37	0.33	0.29	0.26	0.17
Air*	38.89	34.25	30.49	27.43	24.89	22.76	20.99	19.44	18.12	16.90	15.79

*Air vol. fractions: 0.7810 N_2, 0.0092 A, 0.2095 O_2, 0.0003 CO_2.
**Nitrogen + argon.

1.5 COMPUTER PROGRAMS: FLUPROPS, GASSOL

The computer program FLUPROPS contains an interactively acccessible database of fluid properties based on the numerical values presented in Tables 1.3, 1.4 and 1.5. The physical properties of water are calculated by linear interpolation between the relevant tabulated values.

Equation (1.25) is used for the computation of gas viscosity and equation (1.23) for the computation of gas density.

   The computer program GASSOL contains the database on gas solubility presented in Tables 1.6 and 1.7. It provides the user with a computed value for gas solubility at the desired temperature and at gas pressure, inclusive of the aqueous vapour pressure, of 1 atm, on the basis of a linear interpolation between the relevant tabulated values. Listings of these programs are given in Appendix A.

1.5.1   Sample program run: Program FLUPROPS

```
Run
      PROGRAM FLUPROPS

TO ACCESS WATER DATA, ENTER 1
TO ACCESS GASES DATA, ENTER 2

? 1

ENTER THE WATER TEMPERATURE (deg C) ? 22

PHYSICAL PROPERTIES OF WATER AT 22 deg C ARE AS
FOLLOWS:
      DENSITY (kg/m**3) = 997.7704
      DYNAMIC VISCOSITY (Ns/m**2) = 9.5736E-04
      SURFACE TENSION (N/m) = 7.243801E-02
      SATURATION VAPOUR PRESSURE (N/m**2) =
      2669.72

PRESS THE SPACE BAR TO CONTINUE

DO YOU WISH TO OBTAIN FURTHER DATA (Y/N) ? Y

TO ACCESS WATER DATA, ENTER 1
TO ACCESS GASES DATA, ENTER 2

ENTER 1 OR 2, AS APPROPRIATE

? 2

      1. AIR
      2. OXYGEN
      3. NITROGEN
      4. METHANE
      5. CARBON DIOXIDE

SELECT GAS BY TYPING ITS NUMBER ? 2

INPUT GAS TEMPERATURE (deg C) ? 15
INPUT GAS ABSOLUTE PRESSURE (N/m**2) ? 1E5

      DENSITY OF OXYGEN (kg/m**3) = 1.319822
      DYNAMIC VISCOSITY OF OXYGEN (Ns/m**2) =
      2.005136E-5
```

```
              SPECIFIC HEAT AT CONSTANT PRESSURE (J/kg.K) =
              920
              SPECIFIC HEAT AT CONSTANT VOLUME (J/kg.K) =
              657.1
              SPECIFIC GAS CONSTANT (J/kg.K) = 262.9
PRESS THE SPACE BAR TO CONTINUE

DO YOU WISH TO OBTAIN FURTHER DATA (Y/N) ? N

Ok
```

1.5.2 Sample program run: Program GASSOL

```
Run
        PROGRAM GASSOL
```

This program provides solubility values for the following gases in water (gas pressure, inclusive of aqueous vapour pressure, 1 atm):

```
        1. OXYGEN
        2. OZONE
        3. METHANE
        4. NITROGEN
        5. AIR
        6. CARBON DIOXIDE
        7. CHLORINE
        8. CHLORINE DIOXIDE
        9. AMMONIA
        10. HYDROGEN SULPHIDE
        11. SULPHUR DIOXIDE
TO SELECT A GAS, ENTER ITS NUMBER:? 5

ENTER WATER TEMPERATURE (deg C) IN RANGE 0-50: ? 25

SOLUBILITY OF AIR AT 25 deg C IS 22.6 mg/l

Ok
```

REFERENCES

American Public Health Association (1992) *Standard Methods for the Examination of Waters and Wastewaters*, 18th edn, APHA/AWWA/WEF, Washington, USA.

Bernal & Fowler (1933) V. *Chem. Phys.*, **1**, 515.

Camp, T. R. and Meserve, R. L. (1974) *Water and its Impurities*, 2nd edn, Dowden, Hutchinson & Ross, Inc., PA, USA.

Casey, T. J. (1992) *Water and Wastewater Engineering Hydraulics*, Oxford University Press, Oxford, UK.

Crank, J. (1956) *The Mathematics of Diffusion*, Clarendon Press, Oxford, UK.

CRC Handbook of Tables for Applied Engineering Science, 2nd edn (1976) CRC Press Inc., Boca Raton, Florida, USA.

CRC Handbook of Chemistry and Physics, 73rd edn (1992/93) CRC Press Inc., Boca Raton, Florida, USA.

Eisenberg, D. and Kaufman, W. (1969) *The Structure and Properties of Water*, Oxford University Press, Oxford, UK.

Horvath, A. L. (1975) *Physical Properties of Inorganic Compounds*, Edward Arnold Ltd, London, UK.

Maitland, G. C. and Smith, E. B. (1972) *J. Chem. Eng. Data*, **17**, No. 2, 150–155.

Pauling, L. (1953) *General Chemistry*, 2nd edn, W.H. Freeman & Co., San Francisco, USA.

Pauling, L. (1960) *The Nature of Chemical Bond*, 3rd edn, Cornell University Press, Ithaca, New York, USA.

Snoeyink, V. L. and Jenkins, D. (1980) *Water Chemistry*, John Wiley & Sons, New York, USA.

Related reading

Franks, F. ed. (1971–82) *Water. A Comprehensive Treatise*, 7 Vols, Plenum Press, New York.

2

Settling Characteristics of Suspensions

2.1 INTRODUCTION

Waters and wastewaters contain a wide variety of particulate matter, the component particles varying in size, shape and specific gravity. The concentration of particles may also vary widely. Particle concentration may be expressed in volumetric or gravimetric terms.

As discussed in section 1.3.4, the specific gravity of particles is determined by the the density of the particle solid matter and the amount of entrained water. Organic particles, in particular, typically entrain large amounts of water (see section 1.1). In water processes engineering the concentration of particles is mostly expressed in gravimetric terms, i.e. as $mg\,l^{-1}$. However, in particle settling analysis and in the design of particle separation processes, the volumetric concentration ($ml\,l^{-1}$) is also of significance. Where the volumetric concentration is high, the rate of individual particle settling is hindered by the upward movement of water associated with the downward movement of the suspended particles.

In the following sections, three categories of particle settling behaviour are examined: (a) discrete particle settling, where the mutual influence of adjacent particles is negligible; (b) hindered particle settling, which applies to suspensions where the volumetric suspension concentration exceeds a certain threshold value; and (c) so-called 'zone settling' behaviour, which is exhibited by flocculent suspensions that have a sufficiently high volumetric concentration to settle with a well-defined interface between the suspension and the supernatant water.

2.2 SETTLING OF DISCRETE PARTICLES

The forces acting on a discrete particle settling in a water mass are its weight force, F_w, a buoyancy force, F_b, and fluid drag force, F_d. Hence, the rate of change of momentum of a discrete particle of mass m, settling in still water, is defined by these forces as follows:

$$m\frac{\mathrm{d}v}{\mathrm{d}t} = F_w - F_b - F_d \tag{2.1}$$

where

the weight force: $F_w = \rho_s g V$, ρ_s being the particle density and V the particle volume;

the buoyancy force: $F_b = \rho g V$, ρ being the water density;

the drag force: $F_d = C_d A \rho v^2 / 2$, C_d being the drag coefficient, A is the projected particle area in the direction of motion and $\rho v^2 / 2$ is the dynamic or stagnation point pressure.

On inserting these values for F_w, F_b and F_d, equation (2.1) becomes

$$m\frac{dv}{dt} = gV(\rho_s - \rho) - C_d A \rho v^2 / 2 \tag{2.2}$$

The motion of a particle in a centrifuge is subject (neglecting its weight force) to a corresponding set of forces and may be represented by the following equation:

$$m\frac{dv}{dt} = aV(\rho_s - \rho) - C_d A \rho v^2 / 2 \tag{2.3}$$

where a is the radial acceleration $= r\omega^2$, r denoting radius and ω the angular velocity.

As indicated by equation (2.2), it is clear that in gravitational settling, the settling velocity increases until the drag force becomes equal to the submerged weight force. From inspection of equation (2.3) it is clear that, in centrifuging, particle velocity does not converge rapidly to a terminal value because of the fact that a, the angular acceleration, increases with r, the distance of the particle from the centre of rotation.

In gravitational settling (sedimentation), the particle reaches its terminal settling velocity as the accelerating force is reduced to zero:

$$m\frac{dv}{dt} = 0 = gV(\rho_s - \rho) - C_d A \rho v^2 / 2$$

For spherical particles, $A = \pi d^2 / 4$ and $V = \pi d^3 / 6$ and hence the value of the terminal settling velocity v_t is

$$v_t = \left[\frac{1.33gd(\rho_s - \rho)}{C_d \rho}\right]^{0.5} \tag{2.4}$$

The drag coefficient, C_d, depends on the state of flow around the particle, which may be laminar, transitional or turbulent as may be categorised by the particle Reynolds number R_e, which is defined as follows:

$$R_e = \rho d v_t / \mu \tag{2.5}$$

The appropriate R_e value ranges for each type of flow are:

laminar: if $R_e < 1$

transitional: if $1 < R_e < 2000$
turbulent   if $R_e > 2000$

Under laminar conditions $C_d = 24/R_e$, giving the Stokes (1845) expression for terminal settling velocity:

$$v_t = \frac{gd^2}{18\mu}(\rho_s - \rho)$$
(2.6)

or

$$v_t = \frac{gd^2}{18\nu}(S_g - 1)$$
(2.7)

where $S_g = \rho_s/\rho$ is the particle specific gravity and $\nu = \mu/\rho$ is the liquid kinematic viscosity.

Under turbulent conditions C_d is independent of R_e and may be taken to have the value of 0.4 for spherical particles, giving the following expression for turbulent settling velocity:

$$v_t = [3.33gd(S_g - 1)]^{0.5}$$
(2.8)

For the transitional region, Fair and Geyer (1968) suggested the following empirical expression:

$$C_d = 24/R_e + 3/(R_e)^{0.5} + 0.34$$
(2.9)

For all flow conditions other than laminar the drag coefficient is also a function of the shape of the particle and must be determined experimentally. Non-spherical particles will settle more slowly than spherical particles of the same volume and density.

The relationships in the foregoing expressions for the drag coefficient C_d and the terminal settling velocity v_t are summarized graphically in Figures 2.1 and 2.2, respectively.

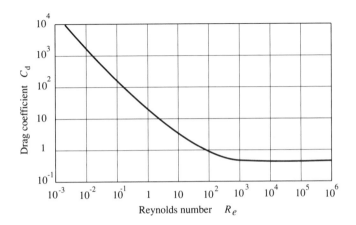

Figure 2.1
Drag coefficient
variation for
spherical particles

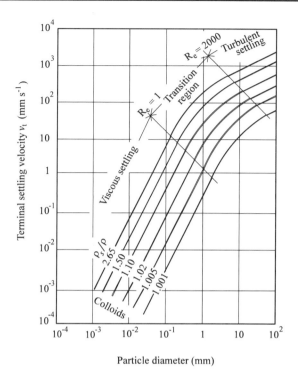

Figure 2.2
Quiescent settling of spherical particles at a water temperature of 10°C

2.3   HINDERED SETTLING OF DISCRETE PARTICLES

In a suspension in which particles are widely separated, the upward flow of displaced water resulting from the downward movement of an individual particle will not affect the rate of settling of neighbouring particles. However, as the *volumetric* concentration of the particles increases, the particles will begin to restrict the area through which the displaced liquid flows upwards, thus hindering downward movement. Settling under such conditions is designated as hindered settling.

The hindered settling velocity of a particle, v_h, may be related to its unhindered settling velocity, v_t, by a relationship of the form

$$v_h = v_t \varphi(c_v) \qquad (2.10)$$

where $\varphi(c_v)$ denotes a function of the volumetric suspension concentration c_v. For organic flocculent suspensions of the type encountered in water and wastewater treatment, the value of c_v is very much influenced by the amount of entrained water in the particles that comprise these suspensions. The volumetric concentration is related to the gravimetric concentration c as follows:

$$c_v = \frac{c}{\rho_s} + \left(\frac{100 - P}{100}\right)\left(\frac{100c}{P\rho}\right) \qquad (2.11)$$

where P is the percentage of dry matter in the particle. Take, for example, an organic particulate suspension where the solid fraction has a specific gravity of 1.4 and the gravimetric concentration is 1000 mg l^{-1} (1 kg m^{-3}). The volumetric concentration is calculated from equation (2.11) to vary with particle water content as follows:

% dry matter:	10	1	0.1
volumetric concentration, c_v:	0.00971	0.09971	0.99971

The following analytical quantification of $\varphi(c_v)$ (Bond, 1960) is based on the flow continuity principle, i.e. it assumes that the particles are settling in an environment in which there is an upward velocity and hence their net settling velocity is equal to their unhindered settling velocity, v_t, minus the upward velocity.

If d is the average dimension of the particles and n is the number of particles in a short length Δl, then in any horizontal area Δl^2, there will be $f_1 n^2 d^2$ area of particles, where f_1 is the shape factor equal to $\pi/4$ for spheres. In any elementary volume Δl^3 there will be $f_2 n^3 d^3$ volume of particles, where f_2 is a shape factor equal to $\pi/6$ for spheres. As the cloud of particles settles, liquid is displaced at a mean upward velocity v_w, such that

$$v_h = v_t - v_w$$

From continuity:

$$v_h f_1 n^2 d^2 = v_w(\Delta l^2 - f_1 n^2 d^2)$$

Combining these two relations:

$$v_h = v_t \left(1 - \frac{f_1 n^2 d^2}{\Delta l^2}\right)$$

since

$$c_v = \frac{f_2 n^3 d^3}{\Delta l^3}$$

therefore

$$v_h = v_t \left(1 - \frac{f_1}{f_2^{2/3}} c_v^{2/3}\right)$$

or

$$v_h = v_t(1 - K c_v^{2/3}) \qquad (2.12)$$

where

$$K = \frac{f_1}{f_2^{2/3}} = 1.21 \text{ for spheres}$$

Bond (1960) found K to have a value of 2.78, on the basis of experiments on alum and lime flocculent suspensions; these suspensions were found to exhibit true hindered settling up to a limiting volumetric concentration of about 0.16. Above this concentration the particles were found to be in partial contact with their neighbours and hence no longer settling as independent entities. The lower c_v limit at which hindering of sedimentation is initiated has been reported to be 0.005 (Camp,1946). As shown above, the gravimetric concentration corresponding to a given volumetric concentration is a function of the specific gravity of the solid matter and the amount of entrained water. Thus, the gravimetric concentration at which hindering of sedimentation is initiated is much lower for porous low-density flocs than it is for impermeable high-density particles.

2.4   ZONE SETTLING

Zone settling is a term used to describe the settling behaviour of highly flocculent suspensions such as activated sludge or the metal hydroxide precipitates produced in chemical coagulation processes. When the concentration of these suspensions exceeds about 500 mg l^{-1}, the floc particles form a loose interconnected mass which settles as a blanket, forming a distinct interface between the settling mass and the supernatant water. Zone settling, as observed in a batch-settling test, is demonstrated in Figure 2.3. The suspended solids column can be divided into three distinct settling zones, the boundaries of which vary as settling proceeds. Initially the solids – liquid interface drops at a uniform rate and constant concentration, exhibiting hindered settling behaviour. This is followed by a decreasing rate of settling with an increasing solids concentration. Finally, the settled solids undergo a gradual compression or compaction.

This type of settling has been analysed by Kynch (1952). In zone-settling behaviour the rate of settling decreases with depth and hence the concentration increases with depth. Kynch assumed that the settling velocity v of a zone-settling particle is a function only of the local

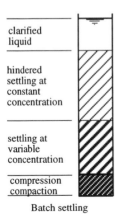

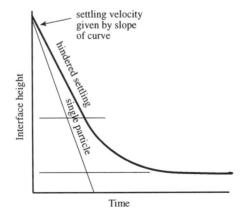

Figure 2.3
Illustration of zone-settling behaviour

gravimetric concentration, $v = f(c)$. Consider any thin layer X within the zone. Let the concentration of solids in this layer be c and let their rate of settling be v. If the rate of settling of solids into this layer is greater than the rate of settling of solids out of it, then its upper boundary must move upwards. Let this upward velocity be u. Let the concentration of solids in the layer immediately above be $(c - dc)$ and the settling velocity of its suspension $(v + dv)$:

inflow of solids to layer X in time $dt = (c - dc)(v + dv + u)A\,dt$
outflow of solids from layer X in time $dt = c(v + u)A\,dt$

Since the concentration in layer X remains constant, inflow must equal outflow:

$$(c - dc)(v + dv + u)A\,dt = c(v + u)A\,dt$$

Neglecting the products of small quantities:

$$u = -v + c\frac{dv}{dc} \tag{2.14}$$

Hence

$$u = -f(c) + cf'(c) = \text{constant}$$

since $v = f(c)$ and c has been assumed as constant. Thus the rate of ascent of any plane of fixed concentration is constant provided the settling velocity of particles within the zone is dependent only on concentration.

The relationship between v and c may be obtained from a batch-settling test. The results of a typical batch-settling test on a suspension exhibiting zone-settling behaviour (e.g. activated sludge) are shown in Figure 2.4. At zero time the suspension has a uniform concentration c_0 and an initial height of h_0. In a settling time t_2, the solids–water interface has dropped to a height h_2; let the concentration at time t_2 be c_2. If A is

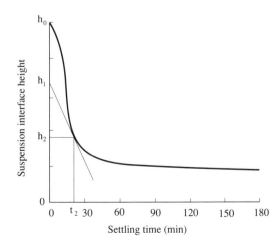

Figure 2.4
Characteristic batch-settling test illustrating the zone-settling behaviour of a flocculent suspension such as activated sludge

the column cross-sectional area, the total mass of solids in the column, $c_0 h_0 A$, will have passed through the plane of concentration c_2 as it ascended from the base of the vessel with velocity u_2. Hence:

$$c_0 h_0 A = c_2 (v_2 + u_2) A t_2 \tag{2.15}$$

The velocity of ascent, u_2, has already been shown to be constant and is therefore equal to h_2/t_2; hence

$$c_0 h_0 = c_2 t_2 \left(v_2 + \frac{h_2}{t_2} \right) \tag{2.16}$$

The settling velocity, v_2, is given by the slope of the curve at the point (h_2, t_2) and from the diagram is seen to be equal to $(h_1 - h_2)/t_2$: hence

$$c_0 h_0 = c_2 t_2 \left[\left(\frac{h_1 - h_2}{t_2} \right) + \frac{h_2}{t_2} \right] \tag{2.17}$$

Therefore

$$c_0 h_0 = c_2 h_1$$

and

$$c_2 = c_0 \frac{(h_0)}{h_1} \tag{2.18}$$

Thus from a batch-settling test the settling velocity corresponding to any given concentration can be computed. It is clear also from the above that h_1 is the height the sludge would occupy if it were uniformly distributed at a concentration c_2.

2.5 SETTLING VELOCITY DISTRIBUTION

The distribution of settling velocity for particles in a *discrete* suspension is determined experimentally in a laboratory settling column test. Laboratory settling columns are usually perspex tubes having a diameter of at least 100 mm and a length of about 2 m, with provision for sampling at various points over the column height. Care is taken to ensure that the suspension is uniformly distributed at the start of a test.

Suppose that samples drawn off at a point 1 m below the water surface at various times gave, on analysis, the results shown in Table 2.1. From this set of results it can be seen that 95% of particles have a

Table 2.1

Time (min)	0	30	60	90	120	180	360
Concentration (mg l^{-1})	42	40	35	29	22	10	2.5
Percentage of initial concentration	100	95	83	69	52	24	6

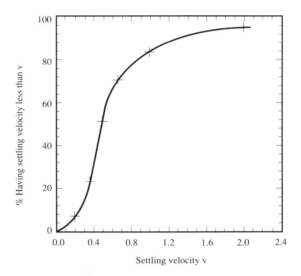

Figure 2.5
Settling velocity
distribution for
discrete particles

settling velocity of less than 1.0 m in 30 min or 2 m h^{-1} and that 83% have a settling velocity of less than 1 m h^{-1}, etc. The cumulative distribution of settling velocities for this suspension is plotted in Figure 2.5. For truly discrete settling, the depth h of the sampling point will not affect the resultant distribution curves of the settling velocities. If the suspension is flocculent, however, a different velocity distribution will be found for each sampling point.

In many practical cases, even in relatively dilute suspensions, particles coalesce to form particle aggregates having increased settling velocities. The extent of such flocculation is a function of many variables including suspension type and concentration, the prevailing velocity gradients, and time. The expected removal of a flocculent suspension in a sedimentation process can be estimated from a laboratory settling test using a suspension column height equal to that used in the process. At various time intervals, samples are withdrawn from ports at different depths and analysed for suspended solids. The percentage removal is computed for each sample and is plotted on a graph against time and depth. Curves of equal percentage removal are then interpolated between the plotted points, as illustrated in Figure 2.6. The resulting curves can be used to determine the overall removal of solids for any detention time and depth within the range of the data, bearing in mind that the test conditions are quiescent. For example, for a detention period equal to t_2 and a depth of h_5 the removal is equal to

$$\frac{\Delta h_1}{h_5} \times \left(\frac{R_1 + R_2}{2}\right) + \frac{\Delta h_2}{h_5} \times \left(\frac{R_2 + R_3}{2}\right) + \frac{\Delta h_3}{h_5} \times \left(\frac{R_3 + R_4}{2}\right) + \frac{\Delta h_4}{h_5} \times \left(\frac{R_4 + R_5}{2}\right)$$

$$(2.19)$$

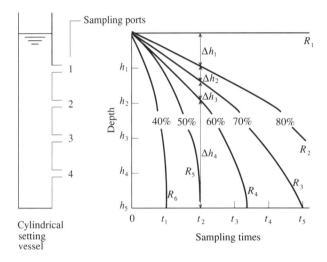

Figure 2.6
Typical settling curves for flocculent suspensions

2.6 NATURE OF SUSPENSIONS IN SEWAGE TREATMENT

The suspensions encountered in sewage treatment include the raw wastewater itself, the humus and activated sludge suspensions produced in biological treatment processes, and the flocculent suspensions generated by physicochemical treatment.

Raw municipal sewage, which is a mixture of domestic and industrial wastewaters, varies in composition from hour to hour, day to day, and location to location. The total solids content of raw sewage (i.e. the sum of the dissolved, colloidal and suspended species of solid matter) depends on the per capita water consumption and the nature of the industrial discharges in the sewer catchment area. It varies in the typical range 500–1000 mg l^{-1}. A typical fractional characterization of the solid matter in sewage as organic, inorganic, dissolved and suspended species is presented in Figure 2.7.

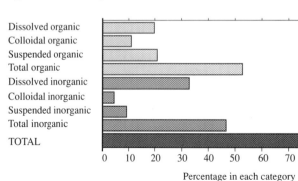

Figure 2.7
Typical proportional distribution of solid matter in raw sewage

The inorganic particulate fraction of municipal wastewater, commonly referred to as *grit*, which typically has a much higher specific gravity and settling velocity than the organic particulate fraction, is conventionally selectively separated in the first stage of treatment. The remaining particulate matter varies in size from 10 mm to colloidal size. Excluding the grit fraction, the specific gravity of the dry solids is in the range 1.0–1.5 but wet solids contain a high proportion of water bringing their specific gravities close to 1.0. Roughly 55–65% of the total suspended solids in municipal wastewaters are settleable and hence removable by a gravity-settling process commonly called primary sedimentation. The nature and volumetric concentration of settleable solids in sewage are such that their settling in primary sedimentation processes may be categorized as hindered flocculent settling. Primary sedimentation generates a solids residue or *primary sludge* which can be readily thickened to achieve a total dry solids concentration in the range 4–8% by weight.

Activated sludge is the microbial suspension or so-called 'mixed liquor' generated in aerobic suspended floc biological treatment processes. It is more consistent in quality and concentration than crude sewage but varies somewhat with operating conditions. The activated sludge particles or flocs range in size from a few microns to more than 1 mm. The specific gravity of the dry solids is over 1.5 but, since this is an organic flocculent suspension containing much entrained water, the in situ specific gravity ranges from 1.01 to 1.03. The mixed liquor suspended solids concentration may vary in the range 1500–5000 mg l^{-1}; being typically highly flocculent, it exhibits the zone-settling behaviour as illustrated in Figure 2.4. However, some activated sludges flocculate poorly and hence have very low settling velocities. This condition is termed 'bulking' and will be discussed further in Chapter 12.

Humus sludge is the microbial biofilm residue generated in aerobic attached film biological processes (aerobic biofiltration). The suspension varies in particle size from colloidal to macroscopic with considerable fluctuation in nature and concentration which are subject to seasonal change. The typical concentration range is 50–150 mg l^{-1} with a specific gravity similar to the organic fraction of crude sewage. The percentage of suspended solids which are setteable is usually in the region of 70–80%.

All these suspensions include colloidal particles which are not amenable to separation by the conventional separation processes. The characteristics of colloidal suspensions and the processes that can be used to effect their destabilization are discussed in Chapter 3.

2.7 NATURE OF SUSPENSIONS IN WATER TREATMENT

The colloidal impurities of natural surface waters are made up of a complex mixture of particles consisting mainly of various colloidal clays. Depending on the degree of contamination, surface waters may also contain colloids from domestic and industrial wastes, live and decaying algae and their decomposition products, bacterial cells, decaying organic matter, and colour colloids. Some of the colloidal impurities, such as clay, may be hydrophobic in nature, whereas others, like certain sewage colloids, may be hydrophilic.

The size of the particles concerned lies between the size of water molecules and the visible matter suspended in the water. Colour colloids are very small and almost approach the size of water molecules. They have been shown by Black and Christman (1963) to be mainly in the 4–10 nm range. Clay particles generally fall in the 0.1–2 μm range.

Because of their small size, these colloidal particles are not visible under an ordinary microscope, even of the highest power. They may be viewed under an electron microscope or by means of the Tyndall effect.

The process of coagulation leads to destabilization and aggregation of colloidal particles into hydroxide floc particles of size up to 1 mm and above, of increased density and settling velocity. These floc particles are of loose structure and contain much entrained water. The coagulating compounds have the following dry densities:

Al oxide	$Al_2O_3(20\,H_2O)$	1180 kg m^{-3}
Fe oxide	$Fe_2O_3(20\,H_2O)$	1340 kg m^{-3}
Crystalline	$Ca\,CO_3$	2600 kg m^{-3}

Their wet densities depend on the amount of entrained water and may calculated using equation (1.16).

Experiments reported by Sterina (1964) give some indication of the effect on the floc of the addition of a long chain polyacrylamide. Artificially turbidified water was used, containing a suspension of clay and floc densities ranging from 1019 kg m^{-3} to 1043 kg m^{-3} were reported. Similar experiments with coloured waters gave values ranging from 1001.5 kg m^{-3} to 1004.7 kg m^{-3}.

With regard to the amount of floc produced, it is reported that the ratio of floc volume to weight of coagulant cation is variable according to the natural water in the range 100–300 ml floc g^{-1} cation. Hudson (1965) found that 1 g Al^{3+} yielded 250 ml floc and 1 g Fe^{3+} produced 250 ml and this value was found independently by Ives (1968). There appears to be no standard method of determining floc volume.

In experiments on an upflow sludge blanket clarifier, Ives and Hale (1970) determined volumetric concentrations in the sludge blanket by lowering a 50 mm diameter tube into the suspension and withdrawing the tube containing the floc from a known location. The floc in the tube was then tested for settlement rate, volumetric concentration and fluidization characteristics. The volumetric concentration was calculated from the volume occupied after 3 h quiescent settlement. It was found that the volumetric concentration varied from 0.20 to 0.05, corresponding to nominal upflow rates ranging from 1 to 2.5 m h^{-1}.

REFERENCES

Black, A. P. and Christman, R. F. (1963) *J. Am. Water Works Assoc.*, **55**, 753.

Bond, A. W. (1960) *Proc. ASCE, Jr. Sanit Eng. Div.*, **86**, SA3, 57–85.

Camp, T. R. (1946) *Trans. ASCE*, **111**, 895.

Fair, G. M., Geyer, J. C. and Okun, D. A. (1968) *Water and Wastewater Engineering* **12**, John Wiley & Sons Inc., New York.

Hudson, H. E. (1965) *Jr. Am. Water Works Assoc.*, **57**, 7, 885.

Ives, K. J. (1968) *Proc. ICE, London*, **39**, 243.

Ives, K. J. and Hale, P. E. (1970) *CIRIA Report 20*.

Kynch, G. J. (1952) *Trans. Faraday Soc.*, **48**, 166.

Sterina, R. M. (1964) *General Report No. 5*, International Water Supply Association Congress, Stockholm.

Stokes, G. C. (1845) *Trans. Cambridge Philos Soc.*, **8**, 287.

3

Destabilization of Colloidal Suspensions

3.1 COLLOIDAL SUSPENSIONS

Many impurities in waters and wastewaters are present as colloidal dispersions, i.e. they occur in particulate form with an approximate size range of 100 nm to 1 nm. Examples of this type of suspension are clays, substances of biological origin such as natural colour, proteins, carbohydrates and their natural or industrial derivatives. Invariably suspensions of this kind possess an inherent stability or resistance to particle aggregation. They are not amenable to clarification by the sedimentation process due to their negligible settling velocity (see Figure 1.6) nor can they be clarified by ordinary filtration processes. Their transformation to a flocculent condition is effected by the process of coagulation.

Colloidal particles have a very high specific surface and consequently their behaviour in suspension is largely determined by surface properties, the gravitational influence on movement being relatively unimportant. Colloidal particles may be hydrophobic (clays, metal oxides, etc.) or hydrophilic (plant and animal residues, proteins, starch, detergents, etc.).

Hydrophobic colloids have no affinity for water and are considered to derive their stability from the possession by individual particles of like charges, which repel each other. These charges may arise from preferential adsorption of a single ion type on the particle surface or from the chemical structure of the particle surface itself. The possession of charge, positive or negative, by a colloid gives rise to its envelopment by an 'electrical double layer', resulting in a potential gradient in the particle vicinity, as shown on Figure 3.1. Since all particles in a given colloidal dispersion are similarly charged, such a suspension is stable by virtue of the electrostatic repelling forces which prevent particles coming together under the influence of Brownian motion and van der Waals attractive forces. It is not possible to measure the potential at the solid-particle boundary but the potential at the rigid-solution boundary or plane of shear can be measured. This latter is called the zeta potential (ζ) and is related to the particle charge and the double layer thickness as follows:

$$\zeta = 4\pi\delta q/D \tag{3.1}$$

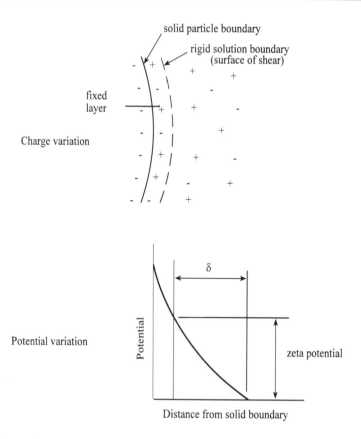

Charge variation

solid particle boundary

rigid solution boundary
(surface of shear)

fixed
layer

Potential variation

δ

Potential

zeta potential

Distance from solid boundary

Figure 3.1
Charge and
potential variation
in the vicinity of
the colloid surface

where δ is the double layer thickness, q is the particle charge and D is the dielectric constant for the liquid medium. The zeta potential is determined experimentally in an externally applied electric field. The ordinary range of the zeta potential is 10 to 200 mV (Fair *et al.*, 1968). Optimum coagulation may be expected to occur when the zeta potential has been reduced to zero (the isoelectric condition of the suspension). For effective coagulation it is necessary to reduce the zeta potential to within 0.5 mV of the isoelectric point (Stumm and Morgan, 1962).

Hydrophilic colloids have, as their name implies, a marked affinity for water. Their stability is due mainly to bound water layers which prevent close contact between particles, although charge is also considered to contribute to their stability. These mainly organic substances may be single macromolecules or aggregates of macromolecules and may be in true solution or in suspension. They derive their charge from the ionization of attached functional groups such as the carboxyl, hydroxyl, sulphato, phosphato and amino groups. The magnitude of this charge is dependent upon the extent of ionization of the functional groups which, in turn, is influenced by the pH of the medium. The charge also influences the solubility of hydrophilic colloids, the minimum solubility being frequently found to coincide with the isoelectric point, which in the majority of cases lies within the pH range 4.0–6.5.

Hydrophilic colloids may become absorbed on hydrophobic colloidal particles such as clays, thereby imparting hydrophilic properties to the latter. Colloidal suspensions of this kind are called 'protective colloids' and may be difficult to coagulate.

3.2 COAGULATION

Coagulation, or the destruction of colloidal stability, may be effected in four major ways: (i) boiling; (ii) freezing; (iii) mutual flocculation by the addition of a colloid of opposite charge; and (iv) the addition of electrolytes. Of these only the latter is of major significance in water and wastewater engineering practice.

Boiling a hydrophobic colloidal suspension may sometimes effect coagulation, mainly through a reduction in the extent of hydration of the colloidal particles and an increase in their kinetic energy. Freezing a colloidal suspension may likewise effect its coagulation by increasing the concentration of the dispersed phase and the ion concentration of the dispersing phase through the growth of pure water ice crystals. Freezing has been proposed as a means of improving the dewatering characteristics of sludges and is also used in the vegetable processing industry for the same purpose. Complete mutual precipitation occurs when colloids of opposite charge are mixed in equivalent amounts in terms of electrostatic charge. This method is not used per se in water and wastewater treatment practice but may occur to some extent in coagulation with iron and aluminium salts.

3.2.1 Effects of electrolytes

Added electrolytes may be considered to act in two ways that tend to reduce the zeta potential and hence the stability, particularly of hydrophobic colloids. By increasing the ionic strength of the dispersing medium they reduce the eletrical double layer thickness, while the adsorption of ions of opposite charge by a colloid reduces its own net charge. This influence of added eletrolyte charge on stability is shown graphically in Figure 3.2. It has been found that the effectiveness of added ions of opposite charge increases dramatically with their valency, a finding first observed by Schulze and Hardy and often expressed as the Schulze–Hardy rule, which states: 'the precipitation of a colloid is effected by that ion of an added electrolyte which has a charge opposite in sign to that of the colloidal particles, and the effect of such ion increases markedly with the number of charges it carries'. The relative coagulating power of a number of electrolytes is shown in Table 3.1.

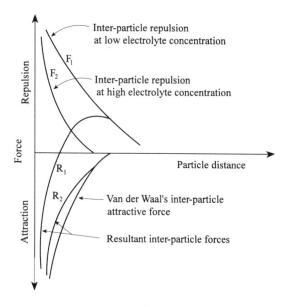

Figure 3.2
Influence of electrolytes on stability (from Fair, 1968. Reproduced by permission of John Wiley & Sons, Inc.)

Table 3.1 Effectiveness of coagulants

Electrolyte	Relative power of coagulation	
	Positive colloids	Negative colloids
NaCl	1	1
Na_2SO_4	30	1
Na_3PO_4	1000	1
$BaCl_2$	1	30
$MgSO_4$	30	30
$AlCl_3$	1	1000
$Al_2(SO_4)_3$	30	>1000
$FeCl_3$	1	1000
$Fe_2(SO_4)_3$	30	>1000

Source: Sawyer and McCarty (1967)

Two other mechanisms also contribute to colloid removal by chemical coagulation. These are (a) enmeshment by the hydroxo-metal precipitate formed by the coagulent chemicals and (b) interparticle bridging. The enmeshment mechanism may be assumed to increase in significance with increasing coagulant dose and hence increasing volumetric concentration of enmeshing solid surface area. The inter-particle bridging mechanism is specifically associated with polyelectrolytes, which are discussed in section 3.4.

The most important coagulating electrolytes in water and wastewater engineering practice are the trivalent salts of aluminium and iron. Their primacy as coagulants is due to (a) their effectiveness (see Table

3.1) in destabilizing the predominantly negatively charged colloids found in natural waters and wastewaters, (b) their low solubility levels in the pH range of normal use (see Figure 3.3) (this is of particular importance in the production of drinking water), and (c) their relatively low cost.

3.3 COAGULATION WITH IRON AND ALUMINIUM SALTS

The coagulating mechanisms of the trivalent salts of iron and aluminium have been widely studied and may be summarized in simplified form as follows: (i) 'Free' trivalent aluminium and iron ions are released in a solution of the respective salts. However, under coagulation conditions only relatively small concentrations of Al^{3+} and Fe^{3+} are present in solution and hence their influence on coagulation may not be as great as is sometimes suggested. (ii) Hydrated metal ions are hydrolysed, producing complex hydrated metal hydroxide ions:

$$Al(H_2O)_6^{3+} \Leftrightarrow Al(H_2O)_5OH^{2+} + H^+ \qquad (3.2)$$

$$Al(H_2O)_5(OH)^{2+} \Leftrightarrow Al(H_2O)_4(OH)_2^+ + H^+ \qquad (3.3)$$

A corresponding set of reactions could be written for trivalent iron. The equilibrium constants for some of these and other similar step reactions are given in Table 3.2. It is important to note that the pH of the solution is depressed by these hydrolytic reactions. Polymerization may also occur:

$$2Al(H_2O)_5(OH)^{2+} \Leftrightarrow Al_2(H_2O)_8(OH)_2^{4+} + 2H_2O \qquad (3.4)$$

It is thought that the extent of polymerization increases with age—the dimer of equation (3.4) may undergo further hydrolytic reactions yielding higher hydroxide complexes, leading to the formation of positively charged colloidal polymers and ultimately to hydroxide precipitates. Both the positively charged complex ions and the colloidal hydroxo polymers are considered to play an important role in the overall mechanisms of coagulation. (iii) The anions of the trivalent salts also play a part in completing the coagulation process by promoting the coagulation of any excess of positively charged metal hydroxo colloids formed. In this respect the greater coagulating power of divalent anions such as sulphates over monovalent anions such as chlorides may be significant.

The most widely used coagulant in drinking water production is aluminium sulphate or alum, having the typical formulation $Al_2(SO_4)_3 18H_2O$. It is a stable, easily handled, readily soluble non-hygroscopic solid. When alum is added to water in the presence of alkalinity the overall reaction is as follows:

$$Al_2(SO_4)_3 18H_2O + 3Ca(HCO_3)_2 \rightarrow 2Al(OH)_3 + 3CaSO_4 + 6CO_2 + 18H_2O$$

$$(3.5)$$

Omitting the non-reacting species, this reaction may be written as

$$Al^{3+} + 3HCO_3^- \rightarrow Al(OH)_3 + 3CO_2 \qquad (3.6)$$

In stoichiometric terms, therefore, $1\ mg\ l^{-1}$ of alum removes $0.45\ mg\ l^{-1}$ alkalinity (as $CaCO_3$) and releases $0.4\ mg\ l^{-1}\ CO_2$. The presence of alkalinity acts as a buffer against excessive lowering of pH, the value of which has an important influence on coagulation. The aluminium hydroxide precipitate formed might be more correctly called a hydrated aluminium oxide precipitate:

$$2Al(OH)_3 \rightarrow Al_2O_3 \cdot 3H_2O \qquad (3.7)$$

It is amphoteric, in that it can react with H^+ or OH^- ions, depending on the pH:

$$Al(OH)_3 + 3H^+ \Leftrightarrow Al^{3+} + 3H_2O \qquad (3.8)$$

$$Al(OH)_3 + OH^- \Leftrightarrow Al(OH)_4^- \qquad (3.9)$$

Table 3.2 Hydrolysis and complex formation equilibria of iron and aluminium

Reaction	Log of equilibrium constant (25 °C)
$Fe^{3+} + H_2O \Leftrightarrow FeOH^{2+} + H^+$	-2.16
$Fe^{3+} + 2H_2O \Leftrightarrow Fe(OH)_2^+ + 2H^+$	-6.74
$Fe(OH)_3(s) \Leftrightarrow Fe^{3+} + 3OH^-$	-38
$Fe^{3+} + 4H_2O \Leftrightarrow Fe(OH)_4^- + 4H^+$	-23
$2Fe^{3+} + 2H_2O \Leftrightarrow Fe_2(OH)_2^{4+} + 2H^+$	-2.85
$Al^{3+} + H_2O \Leftrightarrow Al(OH)^{2+} + H^+$	-5
$7Al^{3+} + 17H_2O \Leftrightarrow Al_7(OH)_{17}^{4+} + 17H^+$	-48.8
$13Al^{3+} + 34H_2O \Leftrightarrow Al_{13}(OH)_{34}^{5+} + 34H^+$	-97.4
$Al(OH)_{3(s)} + OH^- \Leftrightarrow Al(OH)_4^-$	1.3
$2Al^{3+} + 2H_2O \Leftrightarrow Al_2(OH)_2^{4+} + 2H^+$	-6.3
$Al(OH)_{3(s)} \Leftrightarrow Al^{3+} + 3OH^-$	-33

Source: Snoeyink and Jenkins (1980).

Ferric salts (sulphate and chloride) are also widely used as coagulants in water and wastewater treatment. Their reactions in water containing alkalinity are similar to those given above for alum, the net overall precipitation reaction being

$$Fe^{3+} + 3HCO_3^- \rightarrow Fe(OH)_3 + 3CO_2$$

From the foregoing reactions and their equilibrium constants (Table 3.2) it is clear that aluminium and iron are soluble at high and low pH values. The principal ion species remaining in solution are considered to be:

Aluminium: Al^{3+}, $Al(OH)^{2+}$, $Al_7(OH)_{17}^{4+}$, $Al_{13}(OH)_{34}^{5+}$, $Al(OH)^{4-}$, $Al_2(OH)^{24+}$
iron: Fe^{3+}, $FeOH^{2+}$, $Fe(OH)_2^+$, $Fe_2(OH)_2^{4+}$, $Fe(OH)_4^-$

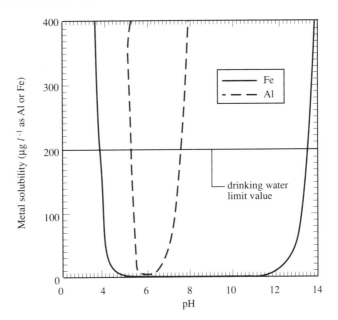

Figure 3.3
Solubility of iron
and aluminium in
water at 25°C

The solubilities of both metals are shown graphically in Figure 3.3 as a function of pH.

The achievement of a low residual metal ion concentration is particularly important in the production of drinking water. The maximum acceptable concentration of both metals is typically set at 200 µg l^{-1} in drinking water standards (WHO, 1993). As may be observed in Figure 3.3, the solubility of iron is less than this limit value over the rather wide pH range 3.7–13.5, while the solubility of aluminium is less than this limit value in the narrower pH range 5.2–7.5. However, it is important to note that these solubility values are equilibrium concentrations and require an equilibrium condition to exist for their realization. This may not always be the case in chemical coagulation practice. It has been observed, for example, that in the chemical coagulation of coloured waters using ferric sulphate, that the residual iron concentrations greatly exceed the solubility values presented in Figure 3.3, especially in the pH range below 5 (Black and Walters, 1964; Haarhoff and Cleasby, 1988; Carroll and Higgins, 1994). The solubility of aluminium in coagulation practice has been found to conform reasonably closely to that presented in Figure 3.3.

Ferrous sulphate, or copperas ($FeSO_4 \cdot 7H_2O$), is also sometimes used as a coagulant on its own or in combination with chlorine in the form of chlorinated copperas:

$$6FeSO_4 + 3Cl_2 \rightarrow 2Fe_2(SO_4)_3 + 2FeCl_3 \qquad (3.11)$$

Other aluminium salts sometimes used in coagulation practice are sodium aluminate, $Na_2O + Al_2O_3 \rightarrow 2NaAlO_2$, which is strongly alkaline in solution and may be used in conjunction with alum, and alumi-nium chlorohydrate, $Al_2Cl(OH)_5$, sometimes used as a dewatering aid for sewage sludges.

3.4 POLYELECTROLYTES

Polyelectrolytes are polymers that contain ionized functional groups such as carboxyl, hydroxyl, amino and other groups. Owing to the presence of these charged sites they possess properties similar to or-dinary low molecular weight electrolytes, e.g. they are soluble and conduct electricity. In solution they may be classified as hydrophilic colloids. Depending on their attached functional groups they may be anionic, cationic or ampholytic. Typical examples of each category are:

$$\text{Polyacrylate} \quad (-CH_2-CH-CH_2-)_n \quad \text{anionic}$$
$$\underset{\displaystyle COO^-}{|}$$

$$\text{Polyvinylpyridinium} \quad (-CH_2-CH-CH_2-)_n \quad \text{cationic}$$

$$N^+$$
$$R$$

$$\text{Polyamino acids} \, (-NH-CH-CO-NH-CH-CO-), \text{ampholytic}$$
$$\underset{(NH_3)^+}{\overset{|}{\underset{(CH_2)_4}{|}}} \qquad \underset{COO^-}{\overset{|}{\underset{(CH_2)_2}{|}}}$$

Activated silica, which is formed by neutralizing a concentrated solu-tion of sodium silicate, $Na_2O + SiO_2 \rightarrow Na_2SiO_3$, to a pH of 6–7, with the consequent formation of polysilicic acid $(H_2SiO_3)_n$, is an anionic polyelectrolyte which has been used as a coagulant aid in water treatment practice, but has largely been displaced by synthetic polymers.

The mechanism of coagulation with polymers is considered to be mainly an adsorptive and bridging one, producing a loose inter-connected network of flocs. Neutralization of charge may also be part of the mechanism, although it has been found that some poly-electrolytes are effective in coagulating colloidal dispersions carrying a charge of the same sign as that possessed by the polyelectrolyte.

In water and wastewater treatment practice polyelectrolytes are nor-mally used in small concentrations as coagulant aids, usually in con-junction with trivalent aluminium or iron salts.

3.5 DETERMINATION OF THE REQUIRED COAGULANT DOSAGE

Because of the many factors that influence coagulation and the complex reactions involved, it is not feasible to calculate directly the coagulant dosage required for coagulation of a particular water or wastewater. An experimental jar test procedure (Cox, 1964) is normally used for this purpose. Using a multiple stirrer apparatus, simultaneous tests are carried out on a series of samples covering a range of coagulant concentration. On addition of the coagulant, the samples are rapidly mixed for 2 min, followed by slow mixing for 20 min. The samples are then allowed to stand for 60 min, after which the colour or turbidity of the supernatant water is measured and the lowest coagulant dose giving adequate removal noted. Using the latter concentration of coagulant a

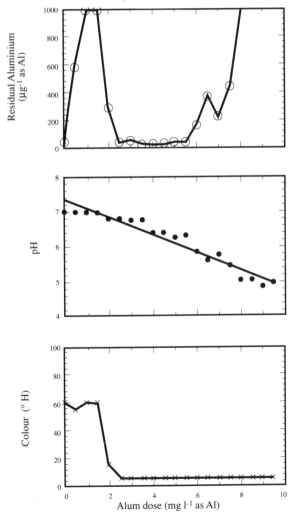

Figure 3.4
Typical jar test results using alum as coagulant (raw water alkalinity 30 mg l^{-1} as CaCO$_3$)

second similar set of tests is performed on pH-adjusted samples to determine the optimum pH value for coagulation. Typical jar test results are shown in Figures 3.4 and 3.5. An alternative but similar jar test procedure, involving the use of the zeta potential measurement in place of the residual colour or turbidity, has also been used.

In water treatment practice the required coagulant dose generally falls within the range $2–8\,mg\,l^{-1}$ as metal ion. In wastewater treatment practice coagulant concentrations up to $40\,mg\,l^{-1}$ (as metal ion) have been used.

3.5.1 Flocculation

When a colloidal suspension has been destabilized, primary floc particles are formed and grow in size through contact with other particles as

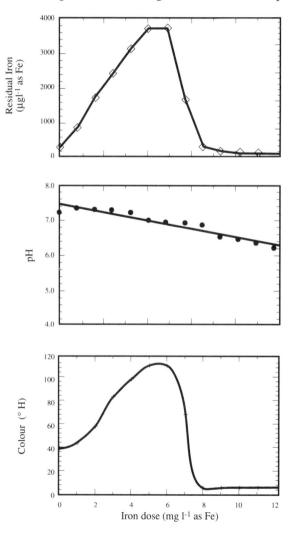

Figure 3.5
Typical jar test results using ferric sulphate (raw water alkalinity $30\,mg\,l^{-1}$ as $CaCO_3$)

a result of Brownian motion. This process is sometimes called peri-kinetic flocculation. As particles grow in size the influence of Brownian effects is diminished and the rate of particle aggregation correspond-ingly reduced. To accelerate the rate of particle collision, velocity gra-dients are created within the body of the dispersing fluid. This controlled use of the velocity gradient to promote flocculation is some-times called orthokinetic flocculation.

3.5.2 The role of the velocity gradient in flocculation

To form aggregates, colloidal particles must be brought sufficiently close together to come under the influence of their van der Waals forces of attraction. Consider any suspended particle having a radius of influ-ence, as determined by the van der Waals force, of R_j. It will attract any other suspended particle i provided the centre of such a particle comes within a distance of $(R_j + R_i)$ of the centre of particle j, i.e. within a sphere of influence of radius $(R_j + R_i)$. Now consider a stream tube of radius $R = R_j + R_i$, with a velocity v at the centre and a constant velocity gradient in the z-direction as shown in Figure 3.6. The volume of fluid within the steam tube, which flows past particle j (located at the centre of the steam tube section) per unit time, is given by

$$\text{vol./time} = 2 \int_0^R z \frac{dv}{dz} b \, dz \qquad (3.12)$$

$$= (4/3)R^3 \frac{dv}{dz} \qquad (3.13)$$

$$= (1/6)(D_i + D_j)^3 \frac{dv}{dz} \qquad (3.14)$$

where D_i and D_j are the diameters of particles i and j, respectively. If there are n_i particles of diameter i per unit volume, then the number of collisions N_j per unit time with particle j will be

$$N_j = (1/6)n_i(D_i + D_j)^3 \frac{dv}{dz} \qquad (3.15)$$

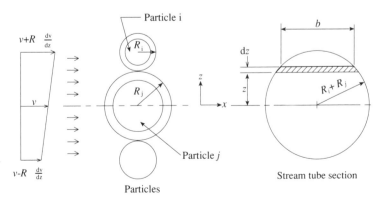

Figure 3.6
Orthokinetic
flocculation

If n_j is the number of particles of diameter j per unit volume, then the number of collisions per unit volume per unit time will be

$$N_{ij} = (1/6)n_i n_j (D_i + D_j)^3 \frac{dv}{dz} \tag{3.16}$$

The continued existence of velocity gradients in a fluid body requires a power input to the fluid. Consider a small cubical element of fluid with flow in the x-direction and a velocity gradient in the z-direction, as shown in Figure 3.7. In conditions of laminar flow the shear stress is linearly related to the velocity gradient:

$$\text{shear stress } \tau = \mu \frac{dv}{dz} \tag{3.17}$$

$$\text{shear force } F = \tau(\Delta x \cdot \Delta y) = \mu \frac{dv}{dz}(\Delta x \cdot \Delta y) \tag{3.18}$$

$$\text{power } P = F \cdot \Delta z \frac{dv}{dz} = \mu \left(\frac{dv}{dz}\right)^2 (\Delta x \cdot \Delta y \cdot \Delta z) \tag{3.19}$$

$$\text{power/volume} = P/V = \mu \left(\frac{dv}{dz}\right)^2 = \mu G^2 \tag{3.20}$$

where

$$G = \frac{dv}{dz}; \text{ the velocity gradient } G = \left(\frac{P}{\mu V}\right)^{1/2} \tag{3.21}$$

The extent of particle aggregation is given by the product of the collision rate and flocculation time, i.e. it is proportional to the product Gt, where t is the flocculation time. Experience suggests that operating Gt values should lie within the range 10^4–10^5. Excessive velocity gradients tend to shear floc particles and must therefore be avoided. The operating range suggested for G is 30–60 s^{-1}.

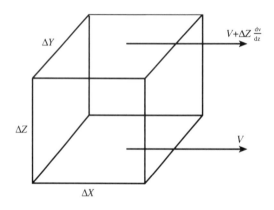

Figure 3.7
Fluid shear strain rate

3.5.3 Mixing techniques and floc growth

Coagulants are normally added to a water or wastewater flow in the form of concentrated solutions. To ensure uniform dispersion of the coagulant a rapid mixing system is required. Following this initial rapid mix a more gentle agitation is required to establish velocity gradients of a magnitude suitable for flocculation. Either a gravitational or mechanical mixing system may be used.

Gravitational systems have the advantages of simplicity and ease of maintenance. Rapid mixing can be achieved by injection into a turbulent pipe flow (velocity > 1 m s^{-1}) or at a hydraulic jump in an open channel.

Horizontal-flow basins with fixed baffles or upward flow sludge blanket type clarifiers may be used to provide the gentle level of mixing required to promote flocculation. Sludge blanket clarifiers, which combine flocculation and floc separation, are described in Chapter 4. They are commonly used in the production of drinking water.

In a horizontal flow baffled basin of volume V, with a flow-through rate Q and head loss h, the power input per unit volume is

$$P/V = \rho g h Q/V \tag{3.22}$$
$$= \rho g h/t \tag{3.23}$$

where t is the detention time. The velocity gradient is

$$G = (\rho g h/\mu t)^{0.5} \tag{3.24}$$

and

$$Gt = (\rho g h t/\mu)^{0.5} \tag{3.25}$$

Baffled basins have the disadvantage that energy dissipation is not uniformly distributed, being excessive at bends and inadequate in the straights. They are also inflexible in operation.

Upward flow sludge blanket clarifiers may be of the hopper-bottom type (Figure 4.10) or the flat-bottom type (Figure 4.11). The sludge blanket is maintained in a fluidized state by the upflow water. The total quantity of sludge in the blanket is kept constant by bleeding off sludge from the top of the blanket at a rate equal to the sludge inflow rate. Velocity gradients are created within the blanket by the shear drag which the blanket exerts on the upward fluid flow. With a stationary blanket, the hindered settling velocity of floc particles at any level must equal the upward bulk flow velocity at that level:

$$v_h = \frac{Q}{A} \tag{3.26}$$

also

$$v_h = v_t(1 - Kc^{2/3}) \tag{3.27}$$

and hence

$$c = \left[\left(1 - \frac{Q}{Av_t}\right)\frac{1}{K}\right]^{3/2} \qquad (3.28)$$

According to equation (3.28), c increases with A. A is constant in flat-bottom tanks, hence c should also be constant over the blanket depth in such tanks. In hopper-bottom tanks c should have a maximum value at the top of the blanket. The magnitude of the velocity gradients created within the blanket can be obtained by equating the drag force on the suspension to its submerged weight:

drag force/unit volume $= c(\rho_s - \rho)g =$ pressure drop per unit length

and power/unit volume $= c(\rho_s - \rho)g\frac{Q}{A} \qquad (3.29)$

$$= \mu G^2$$

and hence

$$G = \left[\frac{c}{\mu}(\rho_s - \rho)g\frac{Q}{A}\right]^{1/2} \qquad (3.30)$$

The value of G is thus theoretically constant in a flat-bottom clarifier. In a hopper-bottom tank, c and A vary with depth. Computations, based on equations (3.27) and (3.30), show (Ives, 1968) that the value of G decreases from the bottom of the blanket to the top in a hopper-bottom tank.

The detention time in the sludge blanket is given by the relation

$$t = V_b(1 - c)/Q \qquad (3.31)$$

where V_b is the sludge blanket volume. The flocculation criterion, Gt, may be calculated as being the product of the mean values of G and t.

Ives (1968) has suggested that the product Gct might be a more significant flocculation criterion than Gt for sludge blanket clarifiers owing to the obvious influence of c on collision opportunity. He did not, however, suggest an optimum value of Gct.

Mechanical mixing systems may be of the high-speed propeller type for rapid mixing or the slow-speed paddle type for flocculation. The power input of the latter may be calculated as follows:

$$P = \frac{C_d A \rho v_r^2 v}{2} \qquad (3.32)$$

where C_d is the drag coefficient for the blades, A is the blade area, v the mean blade velocity and v_r is the velocity of the blade relative to the fluid. Since

$$P/V = \mu G^2$$

then

$$G = \left[\frac{C_d A \rho v_r^2 v}{2V\mu}\right]^{1/2} \qquad (3.33)$$

In flocculation practice (Cox, 1964) peripheral speeds of paddles vary within the range 0.2–0.6 m s^{-1}; the relative blade velocity in the absence of fixed baffles is about 0.75 times the peripheral velocity, and the drag coefficient C_d for flat blades is about 1.8; the paddle blade area is usually between 15% and 25% of tank cross-section area and the detention time is 20–40 min.

3.5.4 Influence of temperature on coagulation

As with virtually all other water or wastewater treatment processes, temperature has a significant influence on coagulation kinetics. A drop in temperature causes: (a) a reduction in the rate of chemical reaction of added coagulants; (b) an increase in water viscosity and hence a reduction in mean velocity gradient for a given power input and hence also a reduction in particle settling velocity; and (c) a reduction in colloid activity. The combined effects of these factors leads to a reduction in process effectiveness with falling temperature.

REFERENCES

Black, A. P. and Walters, J. V. (1964) J. *Am. Water Works Assoc.*, **56**, 99–110.
Carroll, D. E. and Higgins, T. K. (1994) *Aspects of Chemical Coagulation*, Final Year Project Report, Dept of Civil Engineering, University College Dublin.
Cox, C. R. (1964) *Operation and Control of Water Treatment Processes*, WHO, Geneva.
Fair, G. M., Geyer, J. C. and Okun, D. A. (1968) *Water and Wastewater Engineering*, **2**, John Wiley & Sons, Inc. New York.
Haarhoff, J. and Cleasby, J. L. (1988) *J. Am. Water Works Assoc.*, **88**, 168–175.
Ives, K. J. (1968) *Proc. ICE*, **39**, 193.
Sawyer, C. N. and McCarty, P. L. (1967) *Chemistry for Sanitary Engineers*, McGraw-Hill Book Co., New York.
Snoeyink, V. L. and Jenkins, D. (1980) Water Chemistry, John Wiley & Sons Inc., New York.
Stumm, W. and Morgan, J. J. (1962) *J. Am. Water Works Assoc.*, **54**, 971.
WHO (1992) Revision of Guidelines for Drinking Water Quality, WHO, Geneva.

4

Separation of Suspended Solids by Sedimentation Processes

4.1 SETTLING OF DISCRETE PARTICLES IN AN IDEAL FLOW TANK

In an horizontal flow settling tank, the suspension to be clarified flows horizontally through the tank and the suspension settles out on the base of the unit. The classical theory of Hazen (1904) and Camp (1946) is based upon the concept of an ideal horizontal settling zone free from inlet and outlet disturbances in which particles settle freely at their terminal settling velocities as in quiescent conditions. The flow is steady and the velocity of the fluid is everywhere uniform. It is assumed that initially the particles are uniformly distributed across the cross-section of the flow and particles which settle out are not re-entrained. Flocculation, turbulence and other effects are absent.

Referring to Figure 4.1 it can be seen that a particle follows a path determined by the vector sum of its settling velocity v_t and the horizontal velocity v_h. Particles entering the tank at water surface level and

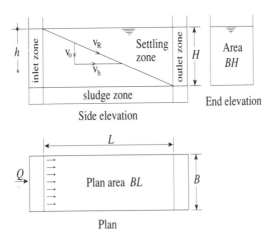

Figure 4.1
Ideal horizontal flow clarifier

having $v_t = v_0$, where $v_0/v_h = H/L$, will just reach the tank floor in the tank flow-through time. Particles having a settling velocity greater than v_0 will also reach floor level, while those particles having settling velocities less than v_0 will be removed from the fluid in the ratio v_t/v_0. Thus all particles having a settling velocity equal to or greater than v_0 will be completely removed, where v_0 is defined by

$$v_0 = v_h \frac{H}{L} = \frac{Q}{BH}\frac{H}{L} = \frac{Q}{BL} \tag{4.1}$$

$Q/BL = Q/A$ is the 'surface loading' or 'overflow rate'. Thus, particles that have a settling velocity equal to or greater than the surface loading, are removed.

The overall removal of particles from a discrete suspension in a horizontal flow rectangular sedimentation basin can be estimated from the suspension settling velocity distribution curve, as illustrated in Figure 4.2. In addition to the removal of particles with settling velocity greater than or equal to v_0, there is a partial removal of particles with settling velocities less than v_0. Referring to Figure 4.2, a proportion v_t/v_0 of the fraction of particles dp, which have a settling velocity v_t, is removed. Hence, the total removal may be written

$$R = (1 - p_0) + \frac{1}{v_0}\int_0^{P_0} v_t \, dp \tag{4.2}$$

The integral is given by the area between the settling curve and the vertical axis and hence the removal ratio for any given value of v_0 (i.e. Q/A) can be computed from the cumulative curve.

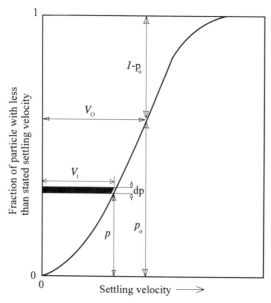

Figure 4.2
Particle settling
velocity distribution

From a theoretical viewpoint, therefore, the removal of discrete particles in horizontal flow settling tanks depends only on surface loading and hence is independent of tank depth and hydraulic retention time. However, in reality the reduction in depth results in a concomitant increase in horizontal velocity resulting in increased turbulence and bottom scour. These effects are discussed later.

The above reasoning may also be applied to an ideal horizontal flow circular tank, where, as in the case of the ideal rectangular tank, it is assumed that the horizontal (radial) velocity is uniform over the tank depth. The radial velocity at radius r, is $v_r = Q/2\pi rH$; while its settling velocity is v_0. Such a particle follows a trajectory defined by the expression

$$\frac{dh}{dr} = \frac{v_0}{v} = \frac{2\pi rHv_0}{Q} \tag{4.3}$$

Integrating this expression between the boundary values of r_1 and r_2, the radii defining the sedimentation zone of the tank;

$$H = \frac{\pi Hv_0}{Q}(r_2^2 - r_1^2) = \frac{Hv_0 A}{Q}$$

or

$$v_0 = \frac{Q}{A}$$

confirming that surface loading is also the key design parameter for circular tanks.

For upward flow sedimentation tanks, it is quite clear that particles having settling velocities less than v_0 will move upwards with the flow and hence are not removed. Thus, referring to the settling velocity distribution shown in Figure 4.2, the theoretical fractional removal of this suspension in a vertical flow settling tank at a surface loading v_0 would be $(1 - P_0)$, where P_0 is the fractional mass of the particles with settling velocities less than v_0.

4.2   RESIDENCE TIME DISTRIBUTION

In the ideal settling tank, the inflow is assumed to be evenly distributed over the cross-section of the tank and the flow is assumed to advance as piston or plug flow to the outlet. Each element of the fluid remains in the tank for a period equal to the theoretical detention time T. However, in an actual tank, the flow pattern is complex because of inlet and outlet disturbances, and because of density, convection and wind-induced currents and dead spaces. Consequently, the hydraulic retention time, instead of being a single value for the flow regime, typically varies over a wide range.

This distribution may be measured by adding a tracer to the inlet flow, as a pulse, and measuring the concentration in the outlet as a function of time. The results of such flow tests are usually plotted as

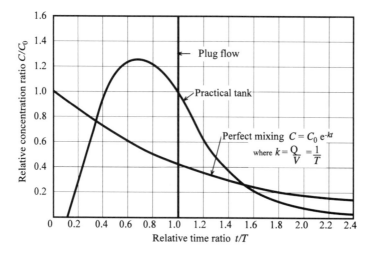

Figure 4.3
Residence time distribution in settling tanks

dimensionless values C/C_0 versus t/T, where C is the effluent tracer concentration at any time t and C_0 is the concentration that would be obtained if the quantity of tracer injected was diluted up to the tank volume. Examples of plug flow, practical tank flow and perfectly mixed flow are shown in Figure 4.3.

It is common to use measures of the central tendency and dispersion of the distribution curve to give qualitative information on the flow pattern in a tank. The mean time, t_{mean}, is defined by the position of the centre of gravity (first moment) of the distribution curve. The modal time, t_{mode}, is the time of flow of the maximum concentration, and the median time, t_{median}, is the time taken for 50% of the added tracer to reach the outlet. Mean, mode and median are all measures of central tendency. If there is no short circuiting their values coincide and the distribution curve is symmetrical, the spread being caused by diffusion alone. The degree of short circuiting may be gauged by the value of

$$\frac{\text{mean-mode}}{\text{mean}} \quad \text{or} \quad \frac{\text{mean-median}}{\text{mean}}$$

The presence of dead space is indicated by the difference between t_{mean} and the theoretical detention time T. In practice, $t_{mean} < T$.

The variability of exposure to sedimentation is indicated by the spread of the curve. A measure of this is the ratio t_{10}/t_{90}, where t_{10} and t_{90} are the times for 10% and 90% of the tracer, respectively, to reach the outlet.

4.3   INFLUENCE OF TURBULENCE

Flow in sedimentation tanks is nearly always turbulent with a consequent reduction in sedimentation due to the lateral movements of settling particles. The Reynolds' number, Re, provides an index of flow conditions:

$$Re = \frac{v_h R}{\nu} \qquad (4.4)$$

where R is the hydraulic radius and ν is the kinematic viscosity. If $Re <$ ca. 600, flow is laminar; if Re is $>$ ca. 2000, flow is turbulent. Re values for rectangular and circular horizontal flow sedimentation tanks may be expressed in terms of flow Q and tank dimensions as follows:

rectangular tanks:

$$Re = \frac{\frac{Q}{BH} \cdot \frac{BH}{B+2H}}{\nu} = \frac{Q}{\nu(B+2H)} \qquad (4.5)$$

circular tanks:

$$Re = \frac{\frac{Q}{2\pi r H} \cdot \frac{2\pi r H}{2\pi r}}{\nu} = \frac{Q}{2\pi r \nu} \qquad (4.5)$$

Thus, to reduce Re for rectangular tanks, the width and/or depth has to be increased. In circular tanks, the flow regime is fixed and Re decreases with radial distance from the centre. The adverse influence of turbulence on sedimentation, based on analyses by Dobbins (1944) and Camp (1946), is shown in Figure 4.4. The extent of reduction of the idealized removal ratio, v_t/v_0, is seen to be a function of v_t/v_h, the ratio of particle settling velocity to horizontal flow velocity.

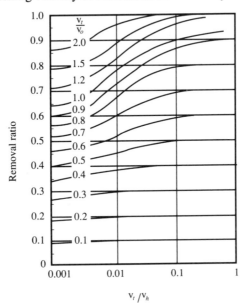

Figure 4.4
Influence of turbulence on the removal ratio in the settling of discrete suspensions. (Reprinted from Camp, 1946 by permission of ASCE.)

4.4 SCOURING OF DEPOSITED PARTICLES

In sedimentation tanks the horizontal flow velocity should be kept below the scour threshold level. For light flocculent particles, such as metal hydroxide flocs and activated sludge, the critical scour velocity (Ingersoll *et al.*, 1956) is defined by the following relationship:

$$\left(\frac{\tau}{\rho}\right)^{1/2} \geqslant v_t \tag{4.7}$$

where τ is the shear stress at the sludge liquid interface and ρ is the supernatant liquid density. $(\tau/\rho)^{1/2}$ is sometimes called the 'shear velocity', generally denoted as v^*.

Considering the flow regime in a rectangular sedimentation basin to be steady open channel flow, the following relationships can be applied to the flow:

$$S_f = \frac{fv_h^2}{8gR_h} \quad \text{and} \quad \tau = \rho g R_h S_f$$

where S_f is the friction slope, v_h is the horizontal velocity $= Q/BH$, f is the friction factor and R_h is the hydraulic radius $= BH/(B+2H)$. It follows from the foregoing flow relationships that

$$\frac{\tau}{\rho} = \frac{f}{8}v_h^2 \tag{4.8}$$

Combining equations (4.7) and (4.8), the criterion for the prevention of scour may be expressed as follows:

$$v_t \geqslant v_h \left(\frac{f}{8}\right)^{1/2}$$

i.e.

$$\frac{Q}{BL} \geqslant \frac{Q}{BH}\left(\frac{f}{8}\right)^{1/2}$$

or

$$\frac{H}{L} \geqslant \left(\frac{f}{8}\right)^{1/2}$$

hence

$$\frac{L}{H} \leqslant \left(\frac{8}{f}\right)^{1/2} \tag{4.9}$$

If f is assumed to have a typical value of 0.024, than the foregoing L/H criterion becomes:

$$\frac{L}{H} \leqslant 18.0$$

To provide a margin of safety in design and to allow for a possible non-uniform velocity distribution, it is advisable to limit the L/H ratio to about 10.

For heavier and more dense particles such as grit, the horizontal velocity v_{sc} required to initiate the scour of deposited particles is given by the following correlation (Camp, 1946):

$$v_{sc} = \sqrt{\frac{8\beta}{f} g (S_g - 1) d} \tag{4.10}$$

where β has a value of about 0.04 for rounded granular material and a value of about 0.06 for non-uniform sticky and flocculent material. For example, the scouring velocity for a grit particle of size 0.2 mm, S_g 2.5, is estimated from equation (4.10), taking $\beta = 0.06$ and $f = 0.025$, to be 0.23 m s^{-1}.

4.5 FLOW STABILITY

Hydrodynamic stability is important in sedimentation in order to reduce the effects on velocity distribution of disturbing influences such as wind, density currents, etc. The flow Froude number Fr can be used as an index of stability:

$$F_r = \frac{v_h^2}{g R_h} \tag{4.11}$$

which, in terms of flow Q and tank dimensions, is expressed as follows:
rectangular tank:

$$Fr = \frac{Q^2}{B^2 H^2} \cdot \frac{1}{g} \cdot \frac{B + 2H}{BH} = \frac{Q^2 (B + 2H)}{g B^3 H^3} \tag{4.12}$$

circular tanks:

$$Fr = \frac{Q^2}{(2\pi r H)^2} \cdot \frac{1}{g} \cdot \frac{2\pi r}{2\pi r H} = \frac{Q^2}{4\pi^2 r^2 H^3 g} \tag{4.13}$$

Thus, flow stability in rectangular tanks is improved by reducing the cross-section area and increasing the length, which is in conflict with the requirements for the reduction of turbulence and the prevention of bottom scour. While there appears to be no generally agreed minimum value of Fr for rectangular tanks, it is considered desirable to have a value in excess of 10^{-5}. In circular tanks, Fr decreases with distance from the tank centre, resulting in poor stability in the outer perimeter zone.

4.6 SEDIMENTATION IN WASTEWATER TREATMENT

Sedimentation is a very widely used solids/liquid separation process in wastewater treatment. It is used in the preliminary stage of treatment for the selective separation of grit, in primary treatment for the separation of settleable solids from the raw wastewater, and in secondary treatment for the separation of biological sludges or chemical precipitates.

4.6.1 Grit separation

The inert fraction of settleable solids in municipal wastewater consisting of ashes, clinker, sand particles etc. is termed grit. It has a higher density than the organic fraction ($S_g > 2$), settles rapidly and, if not removed in the preliminary stage of treatment, would make subsequent sludge handling more difficult. Grit removal units are typically designed to remove grit particles of diameter $\geqslant 0.2$ mm and S_g of 2.6, which have a settling velocity (Figure 2.2) of about 20 mm s^{-1}. Grit separation may be carried out in channel-type grit chambers or in conventional settling basins, known as 'detritors'.

In channel-type grit separators the settlement of putrescible organic solids is prevented by maintaining a scour velocity of 0.3 m s^{-1}. Combining this value with a particle settling velocity of 20 mm s^{-1} results in a required L/H ratio $\geqslant 15$. The horizontal velocity is maintained at a constant value, independent of flow, by an appropriate flow control device at the channel outlet end. Proportional flow (Sutro) weirs (Figure 4.5(a)) are used with channels of rectangular cross-section, while rectangular critical-depth flumes are used with channels of parabolic cross-section (Figure 4.5(b)). These devices, which can also be used for flow measurement, are illustrated in Figure 4.5.

The width x of a Sutro weir decreases with height y (Figure 4.5(a)) according to the following relationship:

$$\frac{x}{b} = 1 - \frac{2}{\pi}\tan^{-1}\left(\frac{y}{a}\right)^{1/2} \tag{4.14}$$

The discharge through a Sutro weir can be expressed in terms of the head h as follows:

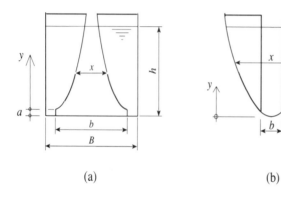

(a)   (b)

Figure 4.5
Grit channel flow control systems. (a) Rectangular channel with proportional-flow weir. (b) Parabolic section channel with critical-depth flume

$$Q = C_d b (2ga)^{1/2} \left(h - \frac{a}{3} \right) \qquad (4.15)$$

Since a is typically small relative to h, Q may be considered to vary linearly with h. Hence the horizontal velocity in the upstream rectangular channel (i.e. Q/Bh) is effectively constant. For a sharp-edged plate weir the discharge coefficient is approximately equal to 0.6. The lower limit value for a and x may be taken as 5 mm.

The geometrical profile of a parabolic channel (Figure 4.5(b)) conforms to the correlation $x^2 = ky$, and its sectional area $A = 2/3xy = 2/3k^{0.5}y^{1.5}$. The discharge through a rectangular critical depth flume may be expressed in terms of the upstream head h as follows:

$$Q = C_d \frac{2}{3} \left(\frac{2}{3} g \right)^{0.5} b h^{1.5} \qquad (4.16)$$

Assuming the discharge coefficient C_d has unit value (rounded upstream converging section), the foregoing expression may be simplified to

$$Q = 1.7 b h^{1.5} \qquad (4.17)$$

Thus the horizontal velocity Q/A through a parabolic channel controlled by a rectangular critical depth outlet flume is

$$\frac{Q}{A} = (1.7 b h^{1.5}) / (\tfrac{2}{3} k^{0.5} h^{1.5}) = \text{constant} \qquad (4.18)$$

'Detritus' tanks of standard rectangular or circular sedimentation tank form, rated at a surface loading of about 20 mm s^{-1}, are also used for grit separation. The grit separated in such tanks must, however, be washed to remove organic material.

The quantity of grit in sewage varies over a wide range (25–250 mg l^{-1}), being greater in discharges from combined sewer systems than in separate systems and also varying with season and rainfall.

4.6.2 Primary sedimentation

The term 'primary sedimentation' in wastewater treatment technology refers to the separation of settleable solids from raw wastewater. The settleable solids in municipal sewage and many organic industrial wastewaters include flocculent and discrete-settling particles. Primary sedimentation tanks at municipal works have to cope with considerable variations in loading, especially when serving combined sewer systems. Primary tanks are typically designed for a maximum hydraulic loading of 3 DWF (dry weather flow). The diurnal flow variation in dry weather is generally in the range 0.5–1.5 DWF. Because of the flocculent nature of the raw sewage solids, solids removal is influenced by retention time as well as by surface loading. The influence of retention time on solids removal is shown in Figure 4.6.

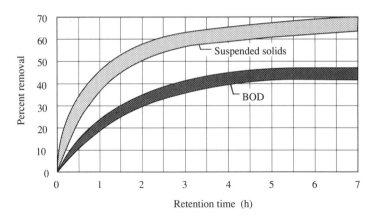

Figure 4.6
Effectiveness of primary sedimentation in the treatment of municipal wastewater.
Source: Fair *et al.*, (1968) (Reprinted by permission of John Wiley, Inc.)

The design surface loading for primary sedimentation tanks is typically in the range 1.0–1.5 m h^{-1} at maximum flow. However, it would appear that significantly higher overflow rates could be used without incurring an excessive penalty in performance, as may be seen from the pilot plant data summarized in Table 4.1. The liquid depth is usually in the range 3.0–4.0 m, giving a retention time of 2–4 h at maximum flow.

Table 4.1   Primary sedimentation pilot plant test results

Surface loading rate ($m^3\,m^{-2}\,d^{-1}$)	Mean influent Sewage solids ($mg\,l^{-1}$)	Sewage solids removal efficiency (%)
25	411	49
50	402	43
100	355	36
150	365	34

Source: Tebbutt and Christoulas 1975.

At small plants, upward flow tanks of hopper-bottom construction, as shown in Figure 4.7, are used. Tanks of this type allow sludge thickening and removal without the aid of mechanical equipment. Because of their shape, however, they are expensive to construct.

Circular tanks, fitted with rotary bridge scrapers, as shown in Figure 4.8, are commonly used in wastewater treatment systems. The scraper blades move the sludge towards a central hopper, from which it is withdrawn. Floor slopes are usually about 1:10. Tanks up to 60 m in diameter have been constructed. The central inflow is directed upwards towards the free surface and its kinetic energy is dissipated within a cylindrical diffusion box or stilling chamber. The diameter of the latter is usually in the range 12%–20% of the tank diameter and typically extends downwards to about mid-tank depth.

Rectangular tanks are also used for primary sedimentation, but usually only at large works. The length:width ratio is typically 3–4:1,

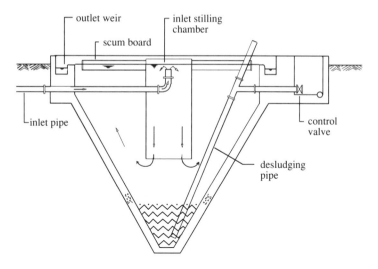

Figure 4.7
Upward flow
clarifier

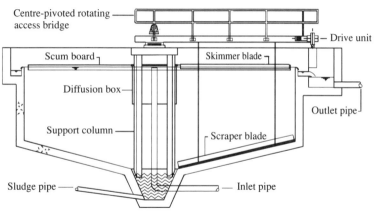

Figure 4.8
Circular clarifier
with rotary bridge
scraper

while the length:depth ratio should preferably not exceed 10 (to avoid scouring effects). Travelling bridge scrapers or submerged flight-type scrapers are used to move settled sludge to hoppers at the inlet end of rectangular tanks. An example of the latter type is shown in Figure 4.9. The floor slope (towards the inlet end) is generally about 1:100.

Primary sludge is preferably withdrawn continuously from sedimentation tanks so as to avoid septicity and the generation of biogas which reduces the effective density of the settled solids and thus disrupts sedimentation. Primary sludge has good thickening characteristics and can be concentrated in primary tank hoppers to a solids content of 4–6%.

4.6.3 Secondary sedimentation

The term 'secondary sedimentation' refers to sedimentation processes following biological (activated sludge, biofiltration) or physicochemical secondary treatment processes. The performance target for secondary sedimentation tanks is generally expressed in terms of a limiting effluent

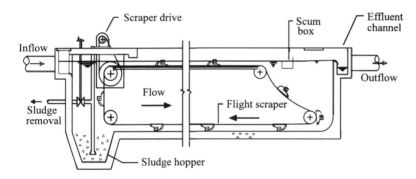

Figure 4.9
Rectangular
primary clarifier
fitted with flight
scraper

solids concentration, typically $\leqslant 30$ mg l^{-1}. The sedimentation basin configurations used in secondary sedimentation applications are generally the same as described for primary sedimentation.

Factors other than those already discussed require to be taken into account in the design of secondary sedimentation basins associated with activated sludge processes. These include the basin solids flux capacity and sludge recycle rate. The activated sludge concentration in aeration basins is typically in the range 2000–4000 mg l^{-1} and thus is significantly higher than other suspensions encountered in wastewater treatment. A design procedure for sizing activated sludge secondary sedimentation basins is presented in Chapter 12. The design surface loading is typically in the range 0.8–1.2 m h^{-1}.

The concentration of suspended solids (humus sludge) in biofilter effluents is characteristically variable but is usually less than 100 mg l^{-1}. Secondary sedimentation basins following biofilters are generally sized on the basis of a design surface loading in the range 1–1.5 m h^{-1}.

The metal hydroxide/organic sludges resulting from physicochemical treatment are flocculent in nature and have settling characteristics generally similar to flocculent activated sludge. They can be separated by conventional sedimentation processes at surface loading rates in the range 0.8–1.2 m h^{-1}.

4.7 SEDIMENTATION IN WATER TREATMENT

Plain sedimentation of source water is rarely necessary in the production of potable water since the settleable solids concentrations in surface waters used as drinking water sources are invariably very low. However, where, river sources intermittently carry large silt loads under flood conditions, it may be necessary to have either intermediate reservoir storage or primary sedimentation as part of the overall treatment system.

Natural surface waters generally contain suspended and colloidal material resulting in colour and turbidity levels not acceptable in drinking water. Dissolved/suspended colour and colloidal solids are converted to a settleable form by the process of chemical coagulation, as described in Chapter 3. The resulting flocculent suspensions are clarified by sedimentation and filtration processes.

Sedimentation tanks of the kind already described for wastewaters may be used for this purpose. However, upward flow fluidized beds or so-called 'sludge-blanket' clarifiers are more usually used for this application.

4.7.1   Sludge-blanket clarifiers

Sludge-blanket clarifiers combine the roles of flocculation and floc separation. As shown in Figures 4.10 and 4.11, the sludge blanket, which has zone-settling characteristics, is fluidized by the upward-flowing water. Sludge is withdrawn from the top blanket, where there is a well-defined interface between the blanket and the supernatant water. Inflow is at floor level, where there is considerable turbulence and mixing which helps to promote flocculation. The earlier form of the sludge-blanket clarifier was of the hopper-bottom type, as shown in Figure 4.10. The hopper-bottom shape has the advantages of a simple inlet system and of providing a simple means of concentrating settled sludge to permit easy removal. Hopper-bottom tanks are, however, expensive to construct; they are limited in plan area if excessive depth is to be avoided, requiring multiple units in large plants. The hydraulic characteristics of these units are also unfavourable owing to the non-uniformity of the divergent upward flow, which causes short-circuiting and blanket instability. For these reasons the surface loading of hopper- bottom sludge-blanket clarifiers is usually limited to 1.0–1.5 m h^{-1}. This range can be extended upwards somewhat by the use of special energy-dissipating inlet devices.

The flat-bottom type of sludge-blanket clarifier typically has a pipe manifold inlet system, located near floor level and discharging down-

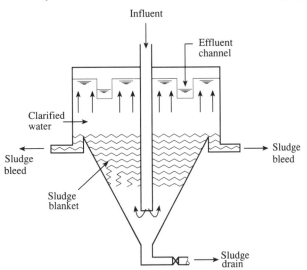

Figure 4.10
Hopper-bottomed sludge-blanket clarifier

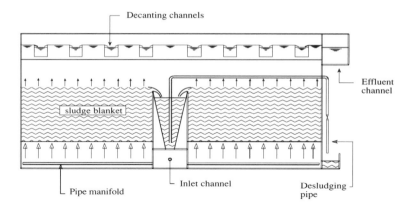

Figure 4.11
Flat-bottomed
sludge-blanket
clarifier

wards. In some designs the flow is pulsed to improve flocculation and prevent the deposition of solids on the tank floor. Clarified effluent is withdrawn by a system of decanting channels at the water surface, as shown in Figure 4.11. To avoid disturbance of the blanket surface by the upward flow, the spacing of the decanting channels is usually not greater than twice the distance between the top of the blanket and the water surface. The design surface loading for flat-bottom sludge-blanket clarifiers is normally in the range 2–4 m h^{-1}, depending on the settling characteristics of the floc.

4.8   INCLINED PLATE AND TUBE SETTLERS

The introduction of inclined plates or tubes into the settling zone of a sedimentation unit, as shown in Figure 4.12, effectively reduces its surface loading. Each pair of plates or individual tube acts as a sedimentation unit. Referring to Figure 4.12, it will be clear that particles having a settling velocity v_0 will be deposited on the plate surface and hence be removed, i.e. v_0 is the effective surface loading. In travelling from A to D, a particle with settling velocity settles through the vertical distance CD. From geometrical considerations, the following relationships are derived:

$$\frac{CD}{v_0} = \frac{AC}{v_1}$$

$$v_0 = v_1\left(\frac{CD}{AC}\right) = v_i\left[\frac{w/\cos\phi}{(H + w/\cos\phi)/\sin\phi}\right]$$

also

$$v_1 \sin\phi = \frac{Q}{A}$$

where Q is the total flow and A is the tank plan area. Hence:

$$v_0 = \frac{Q}{A}\left[\frac{w/\sin\phi}{H/\tan\phi + w/\sin\phi}\right] = \frac{AF}{AE} \qquad (4.19)$$

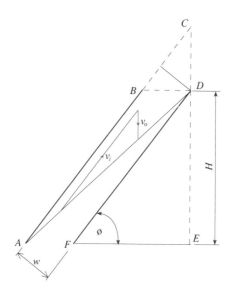

Figure 4.12
Inclined plate pair

Figure 4.12a
Sludge blanket
clarifier fitted with
inclined plates
(Reproduced
courtesy of MC
O'Sullivan &
Company,
Consulting
Engineers, Dublin).

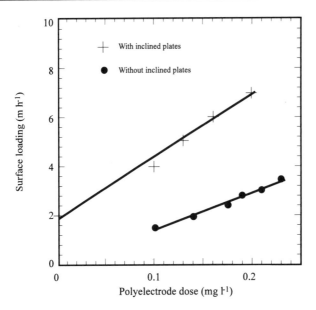

Figure 4.13
Performance of upflow fluidized bed clarifier pilot plant with and without an inclined plate system. *Source:* Purcell, (1983)

Taking, as an example, $w = 0.3$ m, $H = 2.0$ m, and $\phi = 60°$, then

$$v_0 = \frac{Q}{A} \cdot \frac{0.3}{(2 \times 0.5) + 0.3} = 0.23\left(\frac{Q}{A}\right)$$

Thus, the introduction of the plate system as dimensioned has the effect of reducing the surface loading to 23% of the plain tank value. The inter-plate flow is usually laminar, i.e. the inter-plate flow Reynolds number $v_i w / 2\nu < 500$, where ν is the kinematic viscosity.

The practical effect of introducing an inclined plate system into a fluidized bed clarifier of the type illustrated in Figure 4.11 is demonstrated by the experimental results plotted in Figure 4.13. These results were obtained in an upflow fluidized bed clarifier pilot plant at Dublin Corporation's water treatment plant at Ballymore Eustace; the water being treated was an alum-coagulated impounded surface water. The plotted surface loading values are the maximum upflow velocities that could be achieved while maintaining a stable sludge blanket, with and without inclined plates. The plate spacing was 300 m and the plate length was 2.9 m. The results show that the process capacity of the clarifier system was more than doubled by the introduction of an inclined plate system.

4.9   SOME ASPECTS OF HYDRAULIC DESIGN

4.9.1   Inlet systems

While inlet systems vary, depending on the type and shape of the sedimentation tank, the common design goal for all types of inlet system is to achieve a uniform distribution of flow over the flow cross-section. In plain sedimentation processes there is the additional requirement of dissipation of the kinetic energy of the incoming fluid so as not to disrupt the sedimentation process, while in fluidized bed processes there is the rather different additional requirement to inhibit settling of particles on the tank floor.

In rectangular primary tanks, as illustrated in Figure 4.9, the incoming flow is distributed over the tank width by a transverse manifold-type feed channel with multiple uniformly spaced discharge slots or orifices. The kinetic energy from these individual discharges may be locally dissipated by an appropriate baffle system.

In upward flow tanks (Figure 4.7) and circular tanks (Figure 4.8), central stilling chambers are used to contain the turbulent mixing generated by the inflow which, as illustrated in these diagrams, is directed upwards towards the free surface. The stilling chamber diameter is typically (Christie and Harbinson, 1978; Metcalf and Eddy, 1991) within the range 12–20% of the tank diameter and extends from above the water surface to about mid-depth. If the stilling chamber extends too low into the tank, then the inflow may disrupt the settled solids in the central sludge hopper.

In sludge-blanket clarifiers (Figures 4.10 and 4.11), the inflow is directed downward towards the tank floor, thus preventing deposition of solids on the floor. In rectangular flat-bottomed tanks, pipe manifold distribution systems are used, with downward-discharging jets distributing the flow uniformly over the tank floor area. This uniformity of distribution, coupled with a corresponding uniformity of collection by closely spaced surface decanting channels, is essential to the maintenance of a stable sludge blanket in this type of clarifier system.

4.9.2 Outlet systems

It is universal practice to take the outflow from the top of the tank. Weir outlets provide such a simple direct control of tank level that they are almost universally used. Single-sided launders or channels are be preferred to double-sided launders (Ekama and Marais 1986) since it has been found that the approach flow patterns associated with double-sided launders are more likely to cause erosion of the sludge blanket leading to solids loss. Kawamura and Lang (1986) found little difference in clarified effluent quality in a side-by-side comparison of equally loaded twin rectangular tanks, one of which was fitted with in-tank launders to give a weir loading of 10.4 m^3 h^{-1}, while the second tank had a simple transverse end weir resulting in a weir loading of 259.2 m^3 h^{-1}. Thus, there seems little empirical justification for the

Figure 4.14

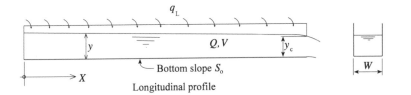

Figure 4.15
Gradually varied
flow in decanting
channel

common design practice of providing in-tank decanting channels to limit the weir loading to a maximum value of about 10 m^3 h^{-1} length of weir.

Plain weirs are unsatisfactory at low overflow rates because small variations in level result in uneven distribution of drawoff. For this reason it is general practice to use notched weir plates on sedimentation tank outlets (Figure 4.14). Circular tanks are typically fitted with a peripheral outlet weir and rectangular tanks with a transverse outlet weir.

Flow in sedimentation tank decanting channels or launders is hydraulically designated as steady gradually varied flow; typically, such channels have a uniform lateral inflow over the channel length and have a free overfall at the outlet end, as illustrated in Figure 4.15. The water surface slope (dy/dx) in such a channel is described by the equation:

$$\frac{dy}{dx} = \frac{S_0 - S_f - \dfrac{Q}{gA^2}\dfrac{dQ}{dX}}{1 - \dfrac{WQ^2}{gA^3}} \tag{4.20}$$

where S_0 is the channel bottom slope, S_f is the friction slope $= fv^2/8gR_h$, Q is the flow, $dQ/dX = q_L$ is the lateral inflow rate per unit length, W is the channel width, and A is the flow cross-sectional area.

Equation (4.20) can be solved numerically, starting from a defined depth at the outlet end. The computer program DECANT calculates the variation in depth, y, along the channel, for channels of circular, rectangular and trapezoidal sections, assuming the depth at the outlet end to be the critical depth y_c, which is defined by

$$\frac{WQ^2}{gA^3} = 1 \tag{4.21}$$

A listing of the program DECANT is given in Appendix A.

4.10 SAMPLE RUN: PROGRAM DECANT

Run

Program DECANT

This program analyses steady gradually varied flow in open channels with uniform lateral inflow over the channel length. The flow depth at the outlet end is assumed to be critical depth, such as would be the case at a free over-fall. The friction head loss is computed using either the Manning or Darcy-Weisbach flow equations, as selected by the user. The program uses a fourth-order Runge-Kutta numerical computational scheme in the solution of the water surface slope equation (4.20), starting from the outlet end where the depth is taken to be 1.02 times the critical depth.

DATA ENTRY

Do you wish to use   1 Manning or   2 Darcy-Weisbach

Enter 1 or 2, as appropriate ? 2

Wall roughness (mm) ? 1

Enter channel data:

Is section   1 CIRCULAR   2 RECTANGULAR
3 TRAPEZOIDAL ?
Enter 1, 2, or 3, as appropriate ? 2

ENTER CHANNEL WIDTH (mm) ? 500

ENTER CHANNEL BED SLOPE ? 0.001

ENTER DISCHARGE (m**3/s) ? 0.16

ENTER CHANNEL LENGTH (m) ? 10

ENTER CHANNEL STEP COMPUTATION LENGTH (m) ? 1

.... data input complete; computation now in progress ....

CRITICAL DEPTH (mm) = 218.6

Distance X (m)	Depth y (mm)	Q (m**3/s)
10	222.9	0.16
9	341.3	0.14
8	350.6	0.13
7	357.7	0.11
6	363.1	0.10
5	367.1	0.08
4	370.0	0.06
3	371.9	0.05
2	372.9	0.03
1	373.0	0.02
0	372.4	0.00

REFERENCES

Camp, T. R. (1946) *Trans. ASCE*, **111**, 895–936

Christie, I. F. and Harbinson, R. W. (1978) *Proc. ICE*, Part 2, **65**, 71–84.

Dobbins, W. E. (1944) *Trans. ASCE*, **109**, No. 2218, 629–656.

Ekama, G. A. and Marais, G. v. R. (1986) Wat. Polut Control, **85**, No. 1, 101–113

Fair, G. M., Geyer, J. C. and Okun, D. A. (1968) Water and Wastewater Engineering, **12**, John Wiley & Sons Inc., New York.

Hazen, A. (1904) *Trans. ASCE*, **53**, Paper no. 980, 45–88.

Ingersoll, A. C., McKee, J. E. and Brooks, N. H. (1956) *Trans. Am. Soc. Civil Eng.*, 1176.

Kawamura, S. and Lang, J. (1986) *J. WPCF*, **58**, No. 12, 1124–1128.

Metcalf and Eddy Inc. (1991) *Wastewater Engineering: Treatment, Disposal, and Re-use*, 3rd edition, McGraw-Hill Publishing Co., Ltd., New York.

Purcell, P. J. (1983) *High-rate sludge blanket clarification*, Thesis presented for the MEngSc degree, Dept. of Civil Engineering, University College Dublin.

Quek, K. H., Bliss, P. J. and Ball, J. E. (1992) *Inst. Eng., Aust.*, **CE34**, No. 1, 37–56.

Tebutt, T. H. and Christoulas, D. G. (1975) *Water Res.* 9, 347–356.

Related reading

Hudson, H. E. Jr (1981) *Water Clarification Processes, Practical Design and Evaluation*, Van Nostrand Rheinhold, New York.

5

Particle Removal by Flotation Processes

5.1 INTRODUCTION

Flotation is a process in which suspensions, the particle phase of which has a specific gravity less than that of the suspending medium, are clarified by allowing the suspended material to float to the surface, where it is removed by skimming. In most applications the effective specific gravity of the suspended phase is artificially lowered by the attachment of gas bubbles. This enables the process to be used for a wide variety of suspended solids, the specific gravities of which are slightly greater than those of their suspending media.

The gas bubbles required to effect the flotation of solids may be generated in a number of ways, including by electrolytic means, by vacuum-activated release of dissolved gases, by air injection through submerged diffusers and by dissolution of air at high pressure in part of the flow, with its subsequent release in fine bubble form on reduction of the pressure to atmospheric level. This latter method of bubble generation is the preferred method in water industry applications; hence the process is generally known as the dissolved air flotation (DAF) process.

The process has long been applied in industry (Gaudin, 1957), especially in mining and refineries, for two-phase separation. It can be used to obvious advantage in water and wastewater treatment systems in place of sedimentation in cases where particle settling velocities are low.

Specific uses of the process are:

(1) to remove paper fibres from pulp and paper mill wastewaters;

(2) to remove oils, greases and similar substances from wastewaters, e.g. food processing, oil refinery and laundry wastes;

(3) to concentrate metallic ores in the mining industry;

(4) to clarify chemically coagulated waters in potable water production; and

(5) to thicken sewage sludge.

5.2   FLOTATION OF SUSPENSIONS

The analysis of the settling velocities of suspensions presented in Chapter 2 is equally valid in the case of suspensions which, because their specific gravity is less than that of the suspending medium, rise rather than settle. The particle settling velocity equations developed in Chapter 2 may therefore also be used to obtain the rise velocity of flotable solids, the equation to be used depending on whether the particle movement is free or hindered. Where the buoyancy of a suspension is artificially increased by the attachment of air bubbles, the situation is somewhat more complex since the bubbles increase in size as they rise to the surface.

The solubility of air in water ($c_{s(a)}$, mg l^{-1}) is linearly proportional to its pressure according to Henry's law:

$$c_{s(a)} = H_a P \qquad (5.1)$$

where H_a is the Henry's law solubility coefficient for air (mg l^{-1} atm^{-1}), which is temperature-dependent, and P is the absolute pressure (atm).

The air solubility relationship defined by equation (5.1) also holds good for most industrial effluents, though the value of H may vary from effluent to effluent and in all cases is less than that for clean water. Data on the saturation concentration of air in clean water at atmospheric pressure, in the temperature range 0–50°C, is presented in Table 1.6 (Chapter 1) and is also plotted on Figure 5.1.

The quantiy of air, c_r, which is released from solution per unit volume of pressurised flow when the pressure is dropped from P (atm) to atmospheric pressure, can be computed from the relation

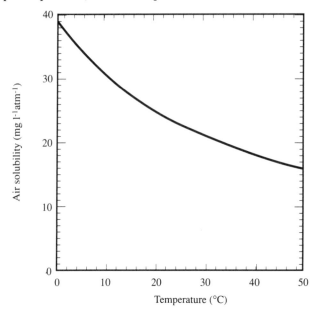

Figure 5.1
Solubility of air in water at a pressure of 1 atm

$$c_r = H_a(Kp - 1) \tag{5.2}$$

where K is the fraction of saturation attained at the elevated absolute pressure P (atm). The actual quantity of air released immediately will be somewhat less than the theoretical value since some time is required to reach an equilibrium condition.

5.3 DETERMINATION OF DESIGN AIR–SOLIDS RATIO

The minimum quantity of air required to float a suspension is that which will reduce the effective density (ρ_c) of the resulting air–particle composite below that of the suspending liquid (ρ):

$$\rho_c = \frac{(V_a \rho_a + V_s \rho_s)}{(V_a + V_s)} \tag{5.3}$$

where V_a is the volume of air attached to a particle of volume V_s, ρ_a and ρ_s being the respective densities of air and particle. The ratio of the associated air and particle masses are

$$\frac{\text{air}}{\text{solid}} = \frac{a}{s} = \frac{V_a \rho_a}{V_s \rho_s} \tag{5.4}$$

Combining equations (5.3) and (5.4) gives

$$\rho_c = \frac{(1 + a/s)}{(1/\rho_s + 1/\rho_a \cdot a/s)} \tag{5.5}$$

Since the composite particle will rise if $\rho_c < \rho$, the minimum a/s required to effect flotation is obtained by setting $\rho_c = \rho$ in equation (5.5), giving

$$\left(\frac{a}{s}\right) \min = \frac{(1 - \rho/\rho_s)}{(\rho/\rho_a - 1)} \tag{5.6}$$

The air–solids ratio is one of the key parameters in the design of flotation systems.

The required air–solids ratio is best determined experimentally using a flotation cell of the type shown in Figure 5.2.

The following procedure may be used:

(1) Partially fill the calibrated cylinder with the suspension to be clarified.

(2) Partially fill the pressure chamber with clarified effluent or clean water depending on which will be used in the flotation system.

(3) Release the flow of compressed air through the pressure chamber to saturate its contents with air at the desired pressure.

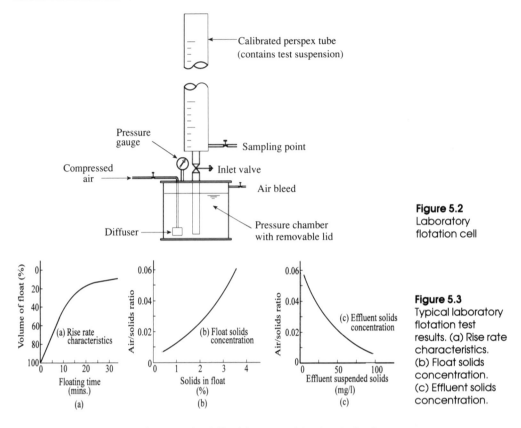

Figure 5.2
Laboratory
flotation cell

Figure 5.3
Typical laboratory
flotation test
results. (a) Rise rate
characteristics.
(b) Float solids
concentration.
(c) Effluent solids
concentration.

(4) Release a volume of pressurized liquid to provide the desired test
air–solids ratio when mixed with the test suspension.

(5) Note the rate of rise of the sludge interface and after a detention
time of 20 min measure the suspended solids content of the clarified
effluent and the floated sludge.

(6) Repeat for various air–solids ratios.

Typical laboratory flotation test results are shown in Figure 5.3.

5.4   INTRODUCTION OF AIR

Air may be introduced to the suspension in the form of finely dispersed
air bubbles released through a diffuser, or the air may be released from
solution in the liquid using vacuum or pressure-drop techniques. In
vacuum flotation the suspension is saturated with air at atmospheric
pressure. This air is released from solution when the suspension is
subjected to vacuum conditions in a flotation cell. The vacuum method
has the disadvantage that the amount of available air is very limited. In
pressure systems part of the clarified effluent, or alternatively a clean

water stream, is partly saturated with air at an elevated pressure (4–6 atm gauge). This high-pressure stream is then mixed with the influent suspension at the flotation tank inlet. Under the influence of a reduction in pressure the excess air is released from the solution in the form of tiny bubbles, which tend to nucleate at the surfaces of suspended particles. The pressure method is the one most used in water engineering practice.

Having selected a suitable a/s ratio the required rate of pressurized flow, Q_p, is obtained as follows:

$$Q_p = \frac{(a/s)Qc}{c_r} \tag{5.7}$$

where c is the influent suspended solids concentration. Insertion of the value of c_r, from equation (5.2) yields:

$$Q_p = \frac{(a/s)Qc}{H_a(Kp - 1)} \tag{5.8}$$

5.5 FLOTATION AIDS

Many of the industrial wastewaters amenable to clarification by flotation are colloidal in nature, e.g. oil emulsions, pulp and paper wastes, and food processing wastes. For the best results such wastes must be coagulated (Chapter 3) before flotation. In metallurgical flotation processes surfactants are often used to depress the liquid surface tension and create a stable foam, which can retain the particles lifted into it.

5.6 DESIGN OF FLOTATION SYSTEMS

The essential elements of a flotation system are an air-saturator and a flotation tank. Schematic layouts are shown in Figure 5.4 (rectangular tank configuration) and Figure 5.5 (circular tank configuration).

Air saturators may be of the packed column type (Casey and Naoum, 1986), as illustrated in Figure 5.4 and 5.5, or may also be of the unpacked column type. The design of these devices is discussed in detail in Chapter 14 (Gas Transfer). They are typically operated in the pressure range 4–6 atm and are designed to achieve up to 90% air saturation. The pipe connecting the saturator to the flotation tank is fitted with a special nozzle or throttle valve (Rykaart and Haarhoff, 1994), which effects the required step-change in pressure between the two units. This sudden drop in pressure across the nozzle results in an almost instantaneous release of the now supersaturated dissolved air in the form of very small bubbles, desirably less than 100 μm in size.

Flotation tanks comprise a mixing or reaction zone, in which the particles and microbubble streams are brought together and particle/

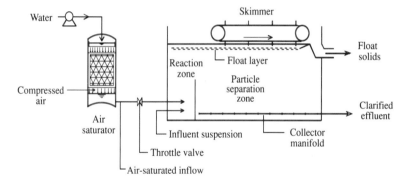

Figure 5.4
Flotation system layout based on rectangular tank

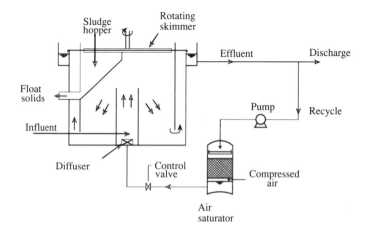

Figure 5.5
Flotation system based on circular flotation tank

bubble agglomerates are formed, and a flotation zone. The reaction zone is confined to a separate compartment ahead of the flotation zone, as illustrated in Figures 5.4 and 5.5. The flotation zone is designed on the same criteria as sedimentation tanks, i.e. on the basis of the rise rate of the suspension and detention time.

The two main applications of the flotation process in water and wastewater engineering are (a) the *clarification* of chemically coagulated waters in drinking water production, particularly for algal-laden waters which are difficult to clarify by sedimentation processes, and (b) the *thickening* of activated sludges.

The design parameter ranges that have been used in practical applications (Haarhoff and van Vuuren, 1993) in the clarification of chemically coagulated surface waters and in the thickening of activated sludges are given in Table 5.1. It will be noted that in Table 5.1 the air quantity required for flotation in water clarification applications is expressed as $mg\,l^{-1}$, rather than as an a/s ratio. Assuming a typical water suspended solids concentration, inclusive of added floculants, of $20\,mg\,l^{-1}$ and an air release quantity of $7\,mg\,l^{-1}$, the corresponding a/s is calculated to be 0.35. If the specific gravity of the solids is taken as 1.5, the theoretical a/s (min) value is calculated from equation (5.6) to be about 0.0004. Thus the theoretical value does not provide a realistic guide for design purposes in this area of application.

Table 5.1 DAF design parameter value ranges

Parameter	Units	Clarification applications	Thickening applications
Reaction zone surface loading	m h^{-1}	40–100	100–200
Reaction zone residence time	min	1–4	0.5–2
Air–solids ratio			0.02–0.04
Air release	mg l^{-1}	6–8	
Crossflow velocity	m h^{-1}	20–100	50–200
Flotation zone surface loading	m h^{-1}	5–11	
Flotation zone solids loading	kg m^{-2}h^{-1}		2–6*, 6–12[†]
Flotation zone side wall depth	m	1.5–3.0	2.0–4.0

* Without coagulants.
[†] With coagulants.
Source: Haarhoff and van Vuuren (1993)

5.7 COMPARISON WITH SEDIMENTATION

The flotation process is particularly suited to the clarification of chemically coagulated surface waters that are low in turbidity and have significant algal concentrations (Janssens and Buekens, 1993). Flotation is not suited to very high turbidity waters (>100 NTU (nephelometric units)). For many water supplies that have a sufficient concentration of natural organic matter that is humic in nature, hydrophobic particles are produced with metal coagulants and there is no need to use polyelectrolytes as coagulant aids (Edzwald, 1994). The hydrophobic nature of the floc particles aids the attachment of air bubbles. The pretreatment flocculation requirements of flotation are different from those of sedimentation. For flotation, it is desirable to produce small strong floc (particle size 10–30 µm), whereas sedimentation requires large particles that will settle quickly.

The advantages of flotation include:

(1) Flotable and settleable solids can be removed in the same unit.

(2) Clarification rates are high, resulting in smaller tank volumes.

(3) A more concentrated sludge is produced.

(4) Oxygenation effects reduce odour problems.

(5) In clarification applications, a smaller floc size is required and hence the use of polyelectrolytes for floc enhancement may not be necessary.

The disadvantages to be noted are:

(1) The quality of effluent produced in terms of effluent suspended solids concentration may not be as good as that obtaind by gravitational settling.

(2) The system is more costly to operate and maintain than gravitational sedimentation systems.

REFERENCES

Casey, T. J. and Naoum, I. E. (1986) *Water Supply*, **4**, 69–82.

Gaudin, A. M. (1957) *Flotation*, 2nd edn, McGraw-Hill Book Co., New York.

Janssens, J. G. and Buekens, A. (1993) *J.* Water Supply Res. Technol—AQUA, **42**, No. 5, 279–288.

Edzwald, J. K. (1994) *Proc. Conf. on Flotation Processes in Water and Sludge Treatment, Proc. IAWQ/IWSA/AWWA, Conf. on Flotation in Water and Sludge Treatment*, Orlando, Florida.

Haarhoff, J. and van Vuuren, L. (1993) *Dissolved Air Flotation, A South African Design Guide*, Water Research Commission, Pretoria, Republic of SA.

Rykaart, E. M. and Haarhoff, J. (1994) *Proc. IAWQ/IWSA/AWWA Conf. on Flotation in Water and Sludge Treatment*, Orlando, Florida.

Related reading

Packham, R. F. and Richards, W. N. (1975) *Water Clarification by Flotation—3*, WRC Tech. Rep. TR 2, Medmenham, Marlow, UK.

Rees, A. J., Rodman, D. J. and Zabel, T. F. (1979) *Water Clarification by Flotation—5*, WRC Tech. Rep. TR 114, Medmenham, Marlow, UK.

Proc. IAWQ/IWSA/AWWA Conf. on Flotation Processes in Water and Sludge Treatment (1994), Orlando, Florida, USA.

Proc. Conf. on Flotation for Water and Wastewater Treatment (1976), WRC, Medmenham, Marlow, UK.

6

Particle Removal by Filtration Processes

6.1 INTRODUCTION

While a great variety of filters are used in water and wastewater treatment practice, they can be differentiated by their mode of action into two broad groups: (a) deep-bed filters and (b) surface filters.

The traditional filter type used in water treatment is the deep-bed filter, consisting of a permeable granular medium (commonly silica sand) through which the water to be filtered flows, and within the pores of which particulate material is retained, hence the description 'deep-bed'.

Surface filters separate particles by a sieving or blocking mechanism; they typically contain a fabric or membrane which permits water flow in response to a pressure difference or gradient but constitutes a barrier to particle transport. The use of membrane filtration in water and wastewater treatment is a more recent development in filtration technology, made possible by the introduction of suitable synthetic membranes in the 1960s and sustained by the ongoing development in membrane technology since that time.

This chapter deals mainly with deep-bed filtration, while the topic of membrane filtration is reviewed briefly in section 6.12.

6.2 DEEP-BED FILTERS

The process of filtering water through beds of granular media in order to purify it is in general use throughout the world. Many different types of filter are used with the principal objective of removing microscopic suspended particles from water. The filters may be broadly classified as 'rapid' or 'slow' according to the rate at which they operate, with the further distinction that in slow sand filtration there is biological activity, whereas in rapid filtration physical removal is the important factor. The process is a dynamic one in which the change in concentration of the suspension flowing through the bed is a function of depth and time.

The two types of filter in common use are known as 'slow' and 'rapid'. They differ both in the rate at which they filter the influent water and in their mode of action. Slow filters treat water at a rate of about 0.1 $m^3 m^{-2} h^{-1}$ (approximately 0.029 $mm s^{-1}$) and rely to a great extent on the formation of a surface biological layer or 'Schmutzdecke' for their cleansing action. In rapid filters the rate of treatment is typically in the range 5–6 $m^3 m^{-2} h^{-1}$ and the full depth contributes to purification and consequently has to be cleaned. Natural silica sand is generally used for both filter types but other media are possible, for example crushed anthracite.

Rapid filters operate at a much higher rate and so require frequent cleaning. This is done by backwashing the sand *in situ* using filtered water, with the aid of mechanical, air or water scouring. The high rate of flow and the need for wash water under pressure adds to the complications of a rapid filter. Sometimes the two systems are combined, a rapid filter being used as a 'roughing' filter to supplement a slow sand plant. The general features of slow and rapid filters are summarized in Table 6.1.

Table 6.1 General features of slow and rapid filters

Parameter	Slow sand filter	Rapid sand filter
Rate of filtration ($m h^{-1}$)	0.1–0.2	5–10
Size of bed (m^2)	5–200	5–200
Depth of sand (m)	0.6–1.2	0.6–1.2
Sand effective size (mm)	0.15–0.3	0.5–1.0
Sand uniformity coefficient	< 3	< 1.5
Grain size distribution	unstratified	stratified by back-washing
Filter floor	designed for filtrate collection only	designed for filtrate collection and back-wash distribution
Loss of head limit (m)	water depth on filter	water depth on filter
Duration of filter run between cleanings (d)	20–60	1–3
Penetration of suspended matter	superficial	deep
Method of cleaning	(1) scraping off surface layer and renewing at intervals (2) mechanical washing of surface layer *in situ*	back-wash with air and water or by water only at high rate
Amount of wash-water used for cleaning sand (% filtrate)	0.2–0.6	2–6
Pre-treatment of water	usually none	coagulation + flocculation + sedimentation or flotation
Supplementary treatment	disinfection	disinfection
Relative capital cost	high	low

6.3   SLOW SAND FILTRATION

Slow sand filters have been used (Graham, 1988) for the purification of public water supplies since the early part of the nineteenth century. In the early days, prior to the introduction of chlorination for water disinfection, slow sand filtration provided an effective barrier against the transmission of waterborne pathogens. For example, Dublin Corporation constructed slow sand filters at Roundwood, Co. Wicklow, in the 1860s to treat impounded River Vartry water. This scheme is still in operation today, supplying about 77 Ml d^{-1}. A schematic slow sand filter layout is shown on Figure 6.1

Slow sand filtration effects a modest removal in colour and can only be used as a sole treatment process where the raw water colour is less than 15–20°H. Where the raw water silt content is high or variable, pretreatment by roughing filters is required. Because of the relatively high colour of many surface waters, slow sand filtration is not widely applicable as a sole treatment process. Worldwide, however, it is still a significant unit treatment process in modern water processes technology, albeit with rather limited application.

A typical slow sand filter unit is composed of 0.15–0.3 mm effective size sand and a uniformity coefficient < 3 (see section 6.8), with a bed thickness in the range 0.6–1 m. The sand bed is unstratified and is supported on a graded gravel layer, designed to prevent the ingress of sand into the underdrain system. The latter, which is designed to convey the filtrate to a central collector channel, may typically consist of a system of perforated lateral pipes discharging to a central manifold pipe or collector channel. Many other types of proprietary underdrain systems are also available, including perforated floors and narrow slit nozzles, the use of which may eliminate the need for the gravel layers. Very small aperture systems should be avoided as they are prone to the development of increased head loss with time due to the

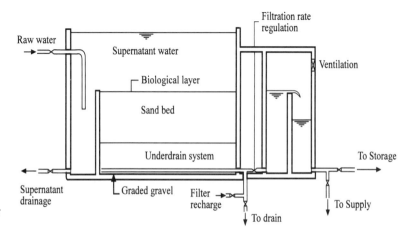

Figure 6.1
Schematic layout of a slow sand filter cell

accretion of material on the material surface. Silica sands and gravels are used.

Biological activity plays an important role in slow sand filtration. After a few weeks operation the topmost layer of sand grains become coated with a gelatinous biological film, which forms a continuous surface mat or 'Schmutzdecke' consisting of bacteria, algae, protozoa and colloidal material derived from the water. This membrane is considered to have a significant barrier-type role in the slow sand filtration process. Bacterial action, resulting in the oxidation of biodegradable substances, extends some distance into the sand bed.

The length of filter run is typically of the order of 30 d. The head loss through the bed, which initially may be about 50–75 mm, increases as the pores in the upper layers of the filter are reduced in size by the deposited material. To avoid negative pressure development within the sand bed (see Figure 6.8) the permissible head loss is usually limited to a value equal to the depth of water above the sand surface, which is typically in the range 1.0–1.5 m. When the head loss has reached the limit value, the bed is drained down and cleaned. The usual method of cleaning is by manual skimming the top 12–20 mm of sand. The filter is then recharged slowly by the upflow of filtered water from storage. The skimmed sand is washed and stored for subsequent re-use. Re-sanding is carried out when the skimming operations have reduced the sand bed thickness to about 0.5 m. At large slow sand filter installations the process of filter cleaning has been mechanized, thus reducing filter downtime. The London Metropolitan Water Board (Rachwal *et al.*, 1988) uses mechanical skimming machines which incorporate an open bottomed box that penetrates 0.15 m into the sand, spanning the width of the filter. The surface is agitated and the sand is cleaned by means of high-pressure water jets which separate the deposited material from the sand grains and remove it from the filter in a washwater stream.

6.4 RAPID FILTERS

Rapid filters are most frequently used in water treatment following pre-treatment by chemical coagulation/clarification; they are also used in wastewater treatment for tertiary effluent 'polishing'. Rapid filters are usually constructed as open-top, free-surface units (rapid gravity filter, RGF) and less frequently as in-line pressure units. Schematic arrangements are shown in Figures 6.2 and 6.3. Pressure units are used mainly in small installations. While the use of pressure filters may eliminate a pumping stage, they have the disadvantage that the condition of the sand bed cannot be visibly inspected.

Rapid filters typically contain about 0.6–1 m of sand overlying an underdrain system. The depth of water overlying the sand surface in RGF units is generally not less than 1.0 m.

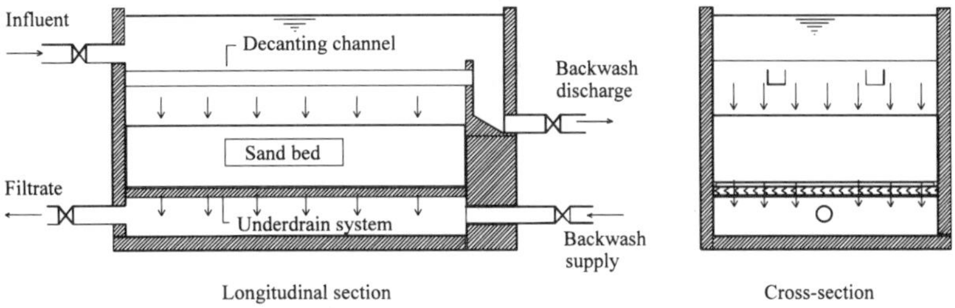

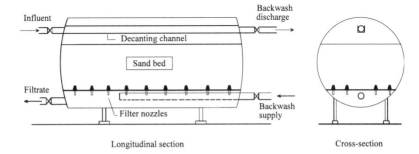

Figure 6.2
Schematic layout
of a rapid gravity
filter

Influent

Decanting channel

Backwash
discharge

Sand bed

Filtrate

Underdrain system

Backwash
supply

Longitudinal section   Cross-section

Backwash
discharge

Influent

Decanting channel

Sand bed

Filtrate

Filter nozzles

Backwash
supply

Figure 6.3
Schematic layout
of a pressure filter

Longitudinal section   Cross-section

6.4.1   Rapid filter underdrain systems

The underdrain system in rapid filters is designed to transmit filtrate
and to ensure a uniform distribution of back-wash water and, where
used, air. It must also prevent loss of sand with the filtrate. Many types
of underdrain system have been devised, two of which are illustrated in
Figure 6.4. The most widely used system consists of a set of perforated
pipe laterals surrounded by graded silica gravel layers, as illustrated in
Figure 6.4(a), which shows a four-layer underdrain system with the
grain size ranging from 2 mm in the uppermost layer to 50 mm in the
bottom layer. The main function of the gravel layers is to distribute
the back-wash upflow evenly over the bed area and prevent the penetration of the sand into the pipe lateral system. Pipe laterals are usually
75–100 mm diameter and are spaced at 150–225 mm centres, with
orifices of 6–12 mm diameter at similar centres, discharging downwards.
They are connected to a central pipe or channel manifold. The manifold
and lateral system are hydraulically designed to ensure a uniform distribution of back-wash water over the filter area. Where air is used for
filter cleaning, a separate air manifold and lateral system is provided.
Alternative underdrain systems incorporating nozzles or no-fines
porous concrete may partly eliminate the need for gravel layers and
may also provide for air distribution, as shown in Figure 6.4(b). In

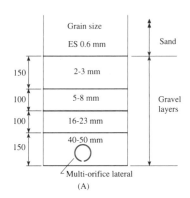

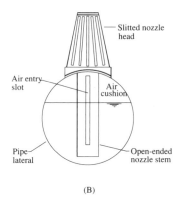

Figure 6.4
Examples of filter underdrainage systems. (a) Graded gravel/perforated pipe lateral system. (b) Slitted nozzle/pipe lateral system

terms of freedom from clogging and long-term performance reliability, the pipe lateral system is probably still the best available system. Casey (1992) presents an analytical procedure for the computation of flow distribution in pipe lateral/manifold systems of the type used in filter underdrain systems and also as flow distribution systems in sludge-blanket clarifiers.

6.5 HYDRAULICS OF FILTRATION

In 1856 Darcy showed that for the steady laminar flow of a liquid through a homogeneous granular material, the rate of flow was linearly related to the pressure gradient:

$$Q = KA\left(\frac{\Delta p}{\Delta l}\right) \tag{6.1}$$

where Q is the volumetric flow rate, A is the cross-sectional area, $\Delta p/\Delta l$ is the pressure gradient and K is the permeability coefficient, the value of which is dependent on the structure of the porous medium and the liquid viscosity. To separate the properties of the liquid from those of the solid material, the permeability coefficient was subsequently modified to

$$K = \frac{k}{\mu}$$

where k is the specific permeability of the medium and μ is the viscosity of the liquid. The specific permeability, k, is an empirical constant which depends on the nature of the material, its mode of packing, porosity and other physical properties. Kozeny (1927) postulated an hydraulic radius model for the flow resistance through a porous medium such as sand. Using the Hagen–Poiseuille equation for laminar flow through a capillary tube:

$$\frac{\Delta h}{\Delta l} = 32 \frac{\mu}{\rho} \cdot \frac{v_c}{g} \cdot \frac{1}{d_c^2} \tag{6.2}$$

where v_c is the mean velocity and d_c is the tube diameter. Expressing d_c in terms of the hydraulic radius, $R = d_c/4$:

$$\frac{\Delta h}{\Delta l} = 2 \cdot \frac{\mu}{\rho} \cdot \frac{v_c}{g} \cdot \frac{1}{R^2} \tag{6.3}$$

The analogous parameter to hydraulic radius (flow area/wetted perimeter) in flow through a sand bed is taken to be the ratio pore volume/grain surface area:

$$R = \frac{\text{pore volume}}{\text{grain surface area}} = \frac{\varepsilon}{(1-\varepsilon)\dfrac{A}{V}}$$

where ε is the porosity, i.e. the ratio of pore volume/total volume, V is the volume of an individual grain and A is the surface area of an individual grain. Equation (6.3) may therefore be adapted to the following form for flow through a uniform sand bed:

$$\frac{\Delta h}{\Delta l} = C \cdot \frac{\mu}{\rho} \cdot \frac{v_c}{g} \cdot \frac{(1-\varepsilon)^2}{\varepsilon^2} \left(\frac{A}{V}\right)^2 \tag{6.4}$$

Expressing the capillary velocity v_c in terms of the approach velocity v, using the correlation $v_c = v/\varepsilon$, yields the Kozeny equation:

$$\frac{\Delta h}{\Delta l} = C \cdot \frac{\mu}{\rho} \cdot \frac{v}{g} \cdot \frac{(1-\varepsilon)^2}{\varepsilon^3} \left(\frac{A}{V}\right)^2 \tag{6.5}$$

where the empirical coefficient C has a value of about 5.

For a spherical sand grain of diameter d_s, A/V has the value $6/d_s$. For a non-spherical sand grain of sieve aperture size d_s, A/V can be expressed as $6/\psi d_s$, where ψ is a sphericity factor, being the ratio of the surface area of the equivalent volume sphere to the actual surface area. Sphericity and bed porosity values for various sand grain shapes are given in Table 6.2.

Table 6.2 Typical porosity and sphericity values for filter sands

Grain description	Sphericity (ψ)	Porosity (ε)
Spherical	1.00	0.38
Rounded	0.98	0.38
Worn	0.94	0.39
Sharp	0.81	0.40
Angular	0.78	0.43
Crushed	0.70	0.48

Assuming a value of 5 for the coefficient C and a sand bed of uniform size spherical sand grains of diameter d_s the Kozeny equation can be written as follows:

$$\frac{h}{l} = 180 \frac{\nu}{g} \frac{(1-\varepsilon)^2}{\varepsilon^3} \frac{v}{(\psi d_s)^2} \tag{6.6}$$

where the kinematic viscosity $\nu = \mu/\rho$. Equation (6.6) is valid while conditions remain laminar, i.e. when the Reynolds number Re is less than 10, where $Re = v d_s/\nu$.

In filter back-washing at high rates the upflow through the sand bed may be outside the laminar flow range. The following empirical equation is suggested for flow in the first part of the transition range, $10 < R_e < 60$:

$$\frac{h}{l} = 130 \frac{\nu^{0.8}}{g} \frac{(1-\varepsilon)^{1.8}}{\varepsilon^3} \frac{v^{1.2}}{(\psi d_s)^{1.8}} \tag{6.7}$$

The maximum head loss in back-washing equals the submerged weight of the sand bed and is reached at incipient fluidization:

$$\left(\frac{h}{l}\right)_f = (1-\epsilon)\left(\frac{\rho_s}{\rho} - 1\right) \tag{6.8}$$

For the typical values of $\varepsilon = 0.4$ and $\rho_s/\rho = 2.6$, $(h/l)_f$ has a value of 0.96, indicating that the maximum head loss through a typical sand bed in upflow is approximately equal to the bed depth. Equation (6.8) can be combined with equation (6.6) or (6.7), as appropriate, to compute the minimum upflow velocity v_{mf} at which fluidization will occur. An increase in upflow velocity beyond this value will not lead to an increase in head loss but will cause the sand bed to expand, increasing its porosity. Wen and Yu (1966) developed the following empirical expression for the expanded bed porosity, in terms of the Reynolds and Galileo numbers:

$$\varepsilon^{4.7} Ga = 18 Re + 2.7 Re^{1.687} \tag{6.9}$$

where Ga is the Galileo number

$$Ga = \frac{d_s^3 \rho(\rho_s - \rho)g}{\mu^2}$$

Wen and Yu also made the following observations:

(1) In multi-size particle systems, the average diameter may be defined by

$$\frac{1}{d_{ave}} = \sum_{i=1}^{n} \frac{X_i}{d_i} \tag{6.10}$$

where X_i is the fraction of particles of size d_i.

Figure 6.5 Layer model of filter bed stratified by backwashing, where d_n = grain size, p_n = fractional depth of layer having grain size d_n and l = bed depth

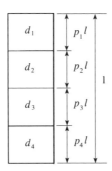

(2) In sand beds of mixed sizes, intermixing occurs if the ratio of sizes is less than 1.3:1, and stratification occurs if it is greater than 1.3:1.

6.5.1 Head loss in a stratified filter

In practical applications filter sands are not of single size but extend over a size range, the size distribution within which can be determined by sieve analysis. The resulting grading curve can be used to approximate the grain size profile within a sand bed stratified by back-washing as consisting of a layered bed, as shown in Figure 6.5.

Equation (6.6) can be applied to each layer, assuming a uniform particle size within a layer:

$$h_n = 180 \cdot \frac{\nu}{g} \cdot \frac{(1 - \varepsilon)^2}{\varepsilon^3} \cdot v \cdot \frac{p_n l}{(\psi d_n)^2} \tag{6.11}$$

where h_n is the head loss across layer n. The total head loss, H, is

$$H = 180 \cdot \frac{\nu}{g} \cdot \frac{(1 - \varepsilon)^2}{\varepsilon^3} \cdot v \cdot l \sum_{i=1}^{n} \frac{p_i}{(\psi d_i)^2} \tag{6.12}$$

This estimation of head loss in a stratified bed assumes that ε and ψ are constant and independent of grain size.

6.6 REMOVAL MECHANISMS IN RAPID FILTRATION

While one might intuitively expect mechanical straining to be a major mechanism of solids separation in rapid filtration, this is not the case. Indeed, such a separation mechanism would rapidly lead to a blinding of the filter surface and a consequent very rapid increase in head loss. In fact, much of the particulate matter removed in rapid filtration is very much smaller than the average filter pore size. The separation of such material results from a combination of transport processes, which bring particles into contact with grain surfaces, and attractive forces, which secure their attachment. The transport mechanisms may include diffusion, hydrodynamic forces, interception and sedimentation, while the attachment mechanisms involve physico-chemical forces.

Diffusion or random Brownian motion is of particular significance for very small particles but has a negligible transport influence compared with gravity and hydrodynamic forces for larger particles. Sedimentation is considered to be significant because of the very large upward-facing horizontal surface area within a filter onto which particles may settle. Removal by sedimentation is a function of particle size and particle density and the velocity distribution within the filter pores.

Inertial forces do not constitute a very significant removal mechanism in rapid filtration because of the low fluid velocities involved and the retarding effect of viscosity.

Surface attachment influences include the universal force of mass attraction (van der Waals force) and electrostic forces which depend on the sign and magnitude of the charges on the particles and on the sand grains (Coulomb force). These forces are only of significance when particles and sand grains are brought close together by one or more of the transport mechanisms noted above. The van der Waals force decreases in proportion to the sixth power of the distance between mass centres, while the Coulomb force decreases in proportion to the second power of the distance between surfaces. Particle attachment is also promoted by the development on the sand grains of a sticky gelatinous layer of deposited material.

6.7 KINETIC ASPECTS OF SUSPENSION REMOVAL IN RAPID FILTRATION

The rate of decrease of suspension concentration c with distance y from the filter bed surface is considered to be proportional to the local concentration in accordance with the following relation:

$$-\frac{dc}{dy} = \lambda c \qquad (6.13)$$

where λ is the filter coefficient (m^{-1}) and is dependent on grain size, porosity and filtration rate. Integration of equation (6.13) subject to the boundary condition of $c = c_o$ at $y = 0$ gives

$$c = c_o e^{-\lambda y} \qquad (6.14)$$

where c is the concentration (mg l^{-1}) of suspended material in the filtrate water at depth y below the filter surface.

The rate of deposition in the filter pores corresponds to the rate of removal from the water. Hence, a mass balance equation may be written for a filter volume of unit plan area and thickness dy as follows:

$$\text{inflow} - \quad \text{outflow} \quad = \text{deposit}$$

$$v \cdot c \cdot dt - v\left(c + \frac{\partial c}{\partial y} dy\right) dt = \frac{\partial \sigma}{\partial t} \cdot dt \cdot dy$$

which simplifies to

$$-\frac{\partial c}{\partial y} = \frac{1}{v} \cdot \frac{\partial \sigma}{\partial t} \tag{6.15}$$

where σ is the mass of deposit per unit volume of filter (specific deposit). Combining equations (6.14) and (6.15):

$$\frac{\partial \sigma}{\partial t} = v \cdot \lambda \cdot c_0 \cdot e^{-\lambda y}$$

Integration subject to the boundary condition $\sigma = 0$ at $t = 0$ yields

$$\sigma = v \cdot \lambda \cdot c_0 \cdot e^{-\lambda y} \cdot t \tag{6.16}$$

Under practical filtration conditions, however, the value of λ is not constant, varying over the filter depth and also varying with time. Its variation with filter depth arises from grain size stratification by back-washing which brings the finer material to the upper regions of the filter. Thus in a clean filter bed λ is likely to decrease with distance from the bed surface. The increase in filter deposited material with the time also causes a decrease in λ, leading to a deeper penetration of deposited material as filtration proceeds.

While much effort has been devoted by many researchers to the development of filtration theory (Camp, 1964; Ives, 1969; Mints, 1966; Burganos et al., 1991 and many others), the development of filtration technology has been largely based on empirical methods and experimentation. This is probably a reflection of the complex nature of the particle removal mechanisms under practical filtration conditions.

6.8   FILTER MEDIUM SELECTION

Silica sand is invariably used as the filter medium in rapid filters. It should be as near single size as possible, otherwise the bed becomes hydraulically stratified by back-washing with a grain size gradient from fine at the top to course at the bottom—the opposite to that required for efficient use of the full depth of the bed. A grain size gradient of course at the top to fine at the bottom can be maintained by using media layers of progressively increasing density, e.g. anthracite on sand on magnetite. The density difference between these materials is sufficient to prevent intermixing during back-washing, provided the size difference is not too great. However, most filters contain sand only because of its ready availability and lower cost than materials such as anthracite and magnetite.

The filter media size distribution may be fully specified, i.e. its grading curve may be required to be within specified upper and lower limits,

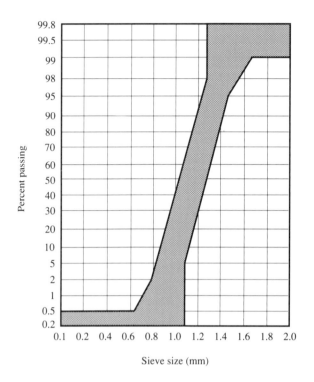

Figure 6.6
Filter sand grading
specification
($d_{10} = 1$ mm;
$d_{60}/d_{10} = 1.2$)

as indicated in Figure 6.6. Alternatively, the size distribution may be specified in terms of the Hazen effective size and uniformity coefficient. The effective size is the sieve size that passes 10% by weight of the material (d_{10}); the uniformity coefficient is the ratio of the sieve size that passes 60% by weight of the material to the sieve size which passes 10% of the material (d_{60}/d_{10}).

Because the smaller grain sizes are of greater significance to filter performance, the ten percentile diameter was chosen as the effective size. The ten percentile by weight corresponds approximately to the fifty percentile by number. Ideally the uniformity coefficient should be as close to unity as possible and preferably less than 1.5. Obviously it is better to set full grading curve limits than to specify an effective size and uniformity coefficient only—the latter specification identifies only two points on the grading curve.

Selection of bed depth and sand size is frequently made on the basis of local experience, an effective size of 0.6 mm and a bed depth of 0.6 m being commonly used. Pilot plant tests provide the best basis for the determination of the design values for these parameters. Pilot units may be constructed in cylindrical tubes of 100–150 mm diameter. After a 'run-in' period, the filter performance is monitored by measuring the head loss at the end of a filter run of the design duration T, while the residual suspended solids concentration (or turbidity level) is measured after a greater duration, e.g 1.1T. These measurements are carried out

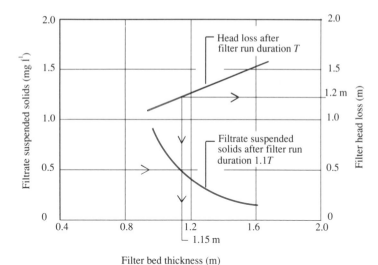

Figure 6.7
Selection of filter
bed thickness on
the basis of pilot
plant test results

for various filter depths and the results are plotted in the form shown in
Figure 6.7. The design depth is then based on a target residual suspended solids or turbidity value, as illustrated in the diagram. The use of
a greater filter run time for the filtrate quality measurement than for the
filter head loss measurement allows the latter to be used for the control
of filter run duration with a built-in margin of safety in relation to
filtrate quality.

Figure 6.7a Backwashing of rapid gravity filter, Ballymore Eustace Waterworks, Dublin. (Reproduced
courtesy of Mr. J. Fenwick, City Engineer, Dublin Corporation).

6.9 BACK-WASHING PRACTICE

Conventional rapid gravity and pressure filters are operated on a batch
mode basis and are taken out of production for back-washing at the end
of each filter run. Back-washing practice varies. The following are
commonly used methods:

(1) Air-scour at an upflow rate of about 30 m h^{-1} for 3 min followed by
    water upflow at a rate of 20 m h^{-1} for 5 min.

(2) Upwash with water only, at a rate of about 50 m h^{-1} for 5 min (used
    in the United States).

(3) Simultaneous air and water upflow for 5 min, air at an upflow rate of
    about 50 m h^{-1} and water at a rate of about 20 m h^{-1}, followed by
    water only at a rate of 20 m h^{-1} for 5 min.

Experimental comparison of these methods (Patterson, 1978) has
shown that method (3), i.e simultaneous use of air and water, provides
the most efficient cleaning action.

The back-washing operation requires a relatively large water flow for
a relatively short period (5–10 min) during which an individual filter
unit is being washed. This can be provided by pumping from low-level
storage or the appropriate volume may be stored at high level to provide
a gravity back-wash supply. The normal sequence of operations in
back-washing is as follows:

(1) Inflow to the filter is shut off.

(2) The standing water level on the filter is drawn down to near bed
    surface level.

(3) The air and water back-wash valves are opened.

(4) After the set duration of the back-washing process, the backwash
    valves are closed.

(5) Filtration is restarted by opening the inflow control.

The back-wash water is collected in decanting channels set about 300
mm above the sand bed surface (sufficiently high so as not to interfere
with bed expansion caused by back-wash). The decanting channels
discharge to waste, as shown in Figure 6.2.

While conventional filters are back-washed intermittently in the
manner just described, innovative filter systems are being developed
which permit continuous filtration and reduce back-wash water
requirements (Boller, 1994).

6.10 CELLULAR SUB-DIVISION OF FILTRATION AREA

The required filter area A is determined by the required filtrate output and the design filtration rate, taking into account the variations in both of these parameters throughout the year. The required water output should include an allowance for the quantity of filtered water to be used for back-washing. The filter area should allow for at least one filter cell being out of use for repair or alteration.

The total filter area A may be divided into n individual units, each of area a in accordance with the relation:

$$a = \frac{A}{(n-1)} \cdot \frac{T}{(T - t_w)} \tag{6.17}$$

where T is the filter run duration and t_w is the filter wash duration. The factor $(n-1)$ allows for one unit being out of commission for repair or alteration; the ratio $T/(T - t_w)$ allows for downtime for back-washing. In general, the value of n should not be less than 3 and the value of a should not be less than 10 m^2. Individual units of very large plan area are undesirable because of the correspondingly large flows required for back-washing—units of plan area greater than 200 m^2 are rare.

6.11 CONTROL OF FILTRATION

The hydraulic control of filtration involves: (1) equal sub-division of incoming flow among the filter units in operation at any time; (2) maintenance of a constant rate of flow through each filter for the full filter run duration; and (3) operation of the back-washing sequence.

A filter run may be terminated when the filtrate quality has deteriorated below a set value, e.g. colour 10°H, turbidity 0.5 FTU or when the head loss has exceeded a set value. Head loss is the preferred control parameter because of its convenience of measurement. A design value of terminal head loss, consistent with the achievement of the set quality target and which also avoids negative head development within the filter, is chosen. As illustrated in Figure 6.8, the development of a negative head can be avoided if the maximum head loss is limited to a value equal to the depth of water above the bed surface.

All filter control systems incorporate an automatic valve on the filtrate line, which gradually opens during a filter run thereby compensating for the increasing filter resistance, so as to keep the filtrate discharge rate constant. A common feed channel with a lateral discharge to each filter unit controlled by a weir or orifice is commonly used to sub-divide flow equally among filter units.

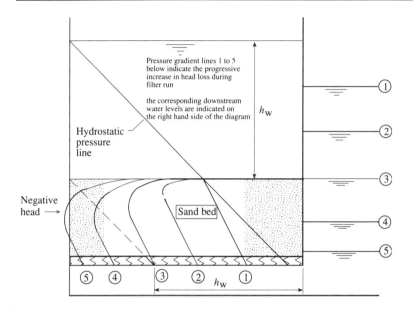

Figure 6.8
Pressure gradient
change during a
filter run

6.12   MEMBRANE FILTRATION

Membrane processes can be broadly classified in two groups on the basis
of the nature of the driving force employed in their operation. The more
widely used membranes are those that rely on an imposed pressure
gradient to force water through the membrane, while retaining particu-
lates and, in some cases, solutes. The second type of membrane process
uses an electrical potential gradient across the membrane to effect the
selective migration of ions in a filtration process known as electrodialysis.

6.12.1   Pressure-driven membrane processes

Pressure-driven membrane processes can be categorized by the molecular weight or particle size cut-off of the membrane, as illustrated in
Figure 1.2 and as summarized in Table 6.3. MF and UF membranes are
considered to effect species separation by a sieving mechanism and
hence they remove particles larger than their size-cutoff limits. The
mode of action of both NF and RO membranes is considered to include
both sieving and diffusion-controlled transport. NF and RO membranes are capable of separating both organic molecules and inorganic
ions. NF and RO membranes are operated at high differential pressure
(up to 100 bar) and have typical characteristic flux rates in the range 5–9 l
$m^{-2} h^{-1} bar^{-1}$. UF membranes are operated at differential pressures in
the range 1–10 bar. MF membranes are operated at differential pressures
$\geqslant 5$ bar and achieve flux rates in the range 100–200 l $m^{-2} h^{-1} bar^{-1}$.

Table 6.3 Classification of pressure-driven membrane separation processes

Membrane process	Size cut-off range (µm)	Examples of materials separated
Microfiltration (MF)	0.05–1.5	microbial cells, large colloids, small particles
Ultrafiltration (UF)	0.002–0.05	macromolecules, viruses, colloids
Nanofiltration (NF)	0.0005–0.007	viruses, humic acids, organic molecules, Ca^{2+}, Mg^{2+}
Reverse osmosis (RO)	0.0001–0.003	aqueous salts, metal ions

There has been an ongoing development in synthetic polymeric membranes since the 1960s which has now reached the stage where membrane filtration can be considered a promising alternative to conventional potable water treatment processes such as chemical coagulation/sand filtration/disinfection. Membranes are marketed in module or cartridge housings, which are supplied in ready-to-use form. They are fitted with feed, permeate and concentrate connections, as schematically illustrated in Figure 6.9. A variety of module configurations are available (Rautenbach and Albrecht, 1989), including tubular, capillary, hollow fibre, spirally wound sheets and plate and frame. Since the permeate discharge per unit membrane area may be relatively low, modules are designed to maximize the membrane packing density, expressed as membrane filtration surface area per unit volume of module (m^2 m^{-3}).

The tubular module consists of a tubular membrane sitting inside a porous sleeve, which in turn fits into a perforated pressure-tight tube between 12 mm and 24 mm diameter. Tubular modules are normally operated in the cross-flow filtration mode.

Capillary modules consist of capillary tube bundles, each end of which is connected to a head plate. The internal capillary diameter is typically in the the range 400–2500 µm and the packing density is of the order of 1000 m^2 m^{-3}. Capillary modules are available in both UF and MF ranges. Like tubular modules, they are normally operated in a cross-flow filtration mode.

Hollow fibre modules consist of hollow fibre bundles, usually bent in hairpin-like fashion, with the open ends cast into an epoxy resin head plate. The internal diameter of the fibres is typically in the range 10–40

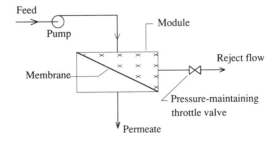

Figure 6.9
Schematic layout
of a membrane
filtration module

μm and the packing density of the order of 10 000 $m^2 m^{-3}$. Unlike the capillary modules, filtration is from the outside inwards, the permeate being discharged through the open fibre ends. Hollow fibre modules are available in the UF, NF and RO ranges.

The spirally wound modules consist of membrane sheets wound around a permeate-collecting tubular core. The rectangular membrane sheets are arranged in 'pockets', each of which consists of a highly porous sheet (permeate-side spacer) sandwiched between two membranes, which are glued together along three edges. The fourth edge of the pocket is attached to the collecting tube. Several such pockets are spirally wound around the collecting tube with a feed-side spacer placed between the pockets to form a so-called 'element'.

In plate and frame modules, the membranes are arranged in parallel flat sheets, with porous spacers, through which the permeate is discharged. The feed solution flows through flat rectangular channels between the membranes. Packing densities are in the range 100–400 $m^2 m^{-3}$.

In general, membrane filters may be operated in either 'dead-end' or 'cross-flow' modes. In the dead-end mode of operation, the permeate is the only outflow from the module during a filter run and hence the removed particulate and solute species accumulate within the module. Cross-flow filtration, on the other hand, operates in a re-circulation mode with the feed introduced tangentially to the membrane surface. The tangential flow promotes the self-cleansing of the membrane surface with an associated reduction in membrane fouling. The suspension can be recycled until the required yield is achieved.

MF processes are used to remove turbidity and chlorine-resistant pathogens such as Cryptosporidium and Giardia. Depending on their cut-off limit, MF membranes may also remove bacteria but do not offer complete removal of viruses, although substantial removal can be achieved if the virus population is largely associated with larger particles (Gallagher et al., 1995). They are typically operated at a differential pressure of less than 1 bar and at flux rates up to 150 l $m^{-2} h^{-1}$.

UF is used in the food and process industries as well as finding increased application in water and wastewater treatment. UF membranes with a molecular weight cut-off of 100 000 were found to achieve a 6 logs rejection (i.e. below the detection limit) of seeded MS2 virus, while UF membranes with larger cut-off limits achieved partial virus removal (Jacangelo et al., 1995). UF hollow fibre membranes are being used in conjunction with powdered activated carbon for the direct production of potable water from surface water (Cornu et al., 1995). Membranes at the lower end of the UF cut-off range can be used to remove humic substances, including colour, from waters (Jensen and Thorsen, 1995).

NF and RO membranes have lower cut-off limits than UF membranes and are capable of rejecting inorganics as well as organic species. NF rejection of inorganics is determined by the composition of the water, the rejection of bivalent ions being significantly greater than the rejection of monovalent ions. RO membranes remove the dissolved organics almost completely and also remove most of the inorganic salts. Hence, they are used for the production of potable water from brackish waters and seawater.

6.12.2 Membrane fouling

Membrane fouling, with a consequent reduction in specific flux, is recognized as a major limitation of the membrane filtration process. It is caused by the gradual build-up on the membrane surface of a fouling layer, which may consist of colloidal particles, iron and/or manganese oxides, and a biofilm, or may be due to inorganic scale-formation, caused, for example, by the precipitation of sparingly soluble salts such as calcium carbonate, magnesium carbonate, calcium sulphate, silica, magnesium hydroxide, etc.

Pre-treatment of the feed water by processes such as chemical coagulation, rapid gravity filtration or some form of cartridge filtration can greatly reduce particulate fouling. Scaling can be eliminated by chemical adjustment of the feed water composition to ensure that the solubilities of the scaling precursors are not exceeded in the retentate (chemical stability is discussed in Chapter 11).

6.12.3 Electrodialysis

Electrodialysis (ED) is a membrane filtration process in which ionic species are removed under the driving force of an electric field. The water to be treated is pumped through a membrane stack which consists of alternately placed anion-permeable and cation-permeable membranes, as illustrated in Figure 6.10. Under the action of the applied electric field the anions migrate towards the anode and the cations migrate towards the cathode, resulting in the stripping of ions from alternate chambers and their concentration in the remaining chambers.

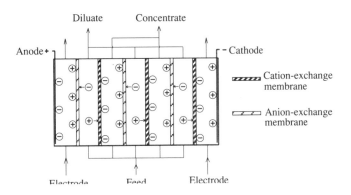

Figure 6.10
General arrangement of electrodialysis module

REFERENCES

Boller, M. (1994) *J. Water Supply Res. Tech.—AQUA*, **43**, No. 2, 65–75.

Burganos, V. N., Paraskeva, C. A. and Payatakes, A. C. (1991) *J. Colloid Interface Sci.*, **148**, 167–181.

Camp, T. R. (1964) *J. San. Eng. Div., ASCE*, **90**, No. SA4, 1–30.

Casey, T. J. (1992) *Water and Wastewater Engineering Hydraulics*, Oxford Science Publications, Oxford University Press, Oxford.

Cornu, S., Gelas, G. and Anselme, C. (1995) *Proc. IWSA Workshop on Membranes in Drinking Water Production*, Paris.

Darcy, H. (1856), *Les fontaines publiques de la Ville de Dijon*, Dalmont, Paris.

Gallagher, P., Taylor, G. S., Hillis, P. R. and Maisuria, J. (1995) *Proc. IWSA Workshop on Membranes in Drinking Water Treatment*, Paris.

Graham, N. J. D. ed., (1988) *Slow Sand Filtration*, Ellis Horwood Ltd, Chichester, UK.

Ives, K. J. (1969) *Proc. IWSA Conf.*, Vienna.

Jacangelo, J., Adham, S. and Laine, J. M. (1995) *Proc. IWSA Workshop on Membranes in Drinking Water Treatment*, Paris.

Jensen, K. and Thorsen, T. (1995) *Proc. IWSA Workshop on Membranes in Drinking Water Treatment*, Paris.

Kozeny, J. (1927) *Sityungsberichte der Wiener Akademic der Wissenschaften*, **136**, Pt. 2a, 271.

Mints, D. M. (1966) *Proc. IWSA Conf.*, Barcelona.

Patterson, P. (1978) *M.Sc. Thesis*, Queen's Univ., Belfast.

Rachwal, A. J., Bauer, M. J. and West, J. T. (1988) Advanced techniques for upgrading large scale slow sand filters, *Slow Sand Filtration*, ed. Graham, N. J. D., Ellis Horwood Ltd, Chichester, UK.

Rautenbach, R. and Albrecht, A. (1989) *Membrane Processes*, John Wiley & Sons, New York.

Wen, C. Y. and Yu, Y. H. (1966) *Mechanics of Fluidisation*, Chem. Eng. Progress, Symposium Series, 62: 62:100.

Related reading

ASCE/AWWA (1990) *Water Treatment Plant Design*, 2nd edn, McGraw-Hill Book Co., New York.

Lorch, W. (1987) *Handbook of Water Purification*, Ellis Horwood Ltd, Chichester, UK.

7

Classification of Dissolved Solids

7.1 CATEGORIES OF DISSOLVED SOLIDS

The number of distinct ionic and molecular species, which constitute the dissolved solids content of natural waters and wastewaters, is quite enormous. In the context of the present discussion on separation principles and related technology, they can be usefully considered under three broad headings: inorganic, organic and dangerous substances.

7.2 INORGANIC DISSOLVED SOLIDS

Inorganic substances in solution may be ionic or molecular. Most inorganics dissociate in solution into their component ions. The predominant ions in natural waters are:

anions: HCO_3^-, Cl^-, SO_4^{2+}, NO_3^-, PO_4^{3-}
cations: Ca^{2+}, Mg^{2+}, Na^+, K^+, Fe^{2+}, Mn^{2+}

Many industrial wastewaters may have high concentrations of inorganics, e.g. wastewaters from fertilizer manufacture, paint manufacuture, electroplating, etc.

Chemical precipitation and ion exchange are the main processes used for the separation of inorganic dissolved solids. For some industrial wastewaters neutralization may be the only treatment required before disposal.

The total inorganic dissolved solids in water can be rapidly estimated instrumentally by conductivity measurement. The total dissolved solids $(mg\, l^{-1})$ in a sample is approximately obtained by multiplying conductivity $(\mu mho\, cm^{-1})$ by an empirical factor. This factor may vary from 0.55 to 0.9, depending on the soluble components of the water and on the temperature of measurement (APHA, AWWA, WEF, 1992). Relatively high factor values apply for saline waters, whereas lower factors apply where considerable hydroxide or free acid is present. The ion concentration $(meq\, l^{-1})$ may be approximated by multiplying the conductivity $(\mu mhos\, cm^{-1})$ by 0.01.

7.3   ORGANIC DISSOLVED SOLIDS

The removal and disposal of dissolved organics constitutes a major part of wastewater engineering practice and is a key measure in water pollution control. Dissolved organics occur in domestic sewage and in wastewaters from three main categories of industrial activity: the processing of natural organic material (meat, vegetables, milk and associated byproducts, etc.); the manufacture of synthetic organic compounds (detergents, pharmaceuticals, by-products of the petroleum industry, etc.); and the fermentation industries (alcohol, organic acids, etc.).

All organic compounds contain carbon in combination with one or more elements. The hydrocarbons contain only carbon and hydrogen. Carbon, hydrogen and oxygen are the major elements of a great number of organic compounds. Minor elements in naturally occurring organic compounds include nitrogen, phosphorus and sulphur. Synthetic organic compounds may contain, in addition, halogens, certain metals and a wide variety of other elements (Sawyer and McCarty, 1967).

Organic compounds differ from inorganic substances in a number of respects, including:

(1) Organic compounds are generally less soluble in water.

(2) The reactions of organic compounds are usually molecular rather than ionic and hence are often quite slow.

(3) The same chemical formula may represent several organic compounds—this is called isomerism.

(4) Many organic compounds have a high molecular weight, often over 1000.

(5) Most organic substances are biodegradable.

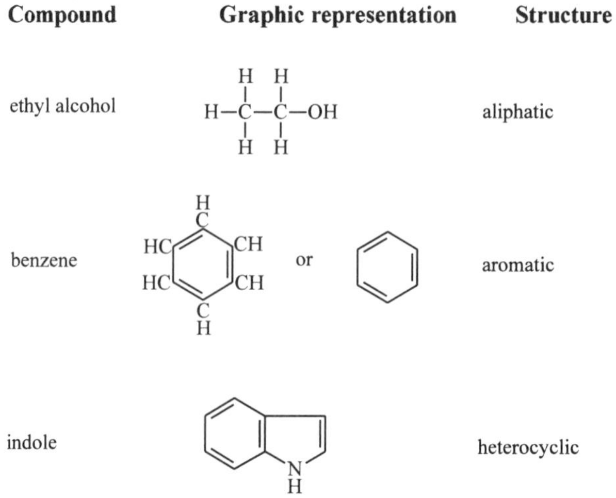

Figure 7.1

Organic compounds can be categorized into three major groupings on the basis of the molecular arrangement of the carbon atoms. These are the *aliphatic, aromatic* and *heterocyclic* compounds. The aliphatic compounds are characterized by a straight or branched carbon chain. The aromatic compounds have a repeated ring structure, each ring consisting of six carbon atoms and containing three double bonds. The heterocyclic compounds have a ring structure in which one member is an element other than carbon. The molecular structure of these groupings is illustrated in Figure 7.1, while a listing of organic compound categories of sanitary significance is given in Table 7.1.

Table 7.1 Major groupings of organic compounds

Aliphatic compounds	Aromatic compounds	Heterocyclic compounds
Hydrocarbons	Hydrocarbons	Furaldehyde
Alcohols	Phenols	Pyrrole
Aldehydes	Alcohols	Pyrrolidine
Ketones	Aldehydes	Indole
Organic acids	Ketones	Skatole
Esters	Organic acids	
Ethers	Amines	
Alkyl halides	Nitrobenzenes	
Polyhalogen compounds		
Amines		
Amides		
Nitriles		
Mercaptans (thioalcohols)		

Naturally occurring organic compounds fall into three major groupings: *carbohydrates, fats* and *proteins.*

The designation carbohydrate is applied to a large group of organic compounds, having the general atomic composition $(CH_2O)_n$, i.e. the number of hydrogen atoms is twice the number of oxygen atoms, as in water. Carbohydrates are grouped into three general classifications: (1) simple sugars or monosaccharides; (2) complex sugars or disaccharides; (3) polysaccharides (Figures 7.2–7.4).

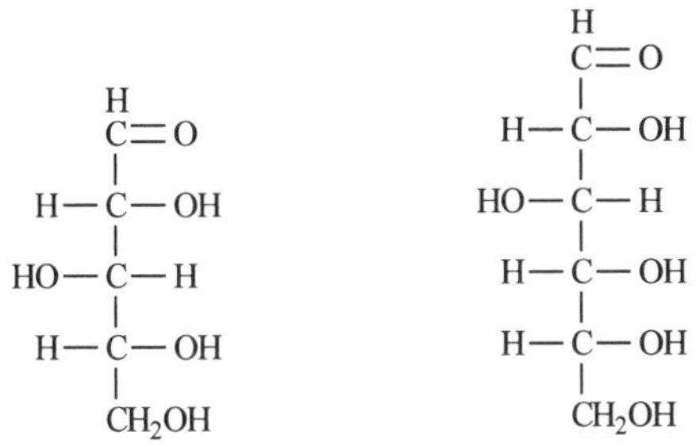

Figure 7.2
Simple sugars

Xylose (pentose)   Glucose (hexose)

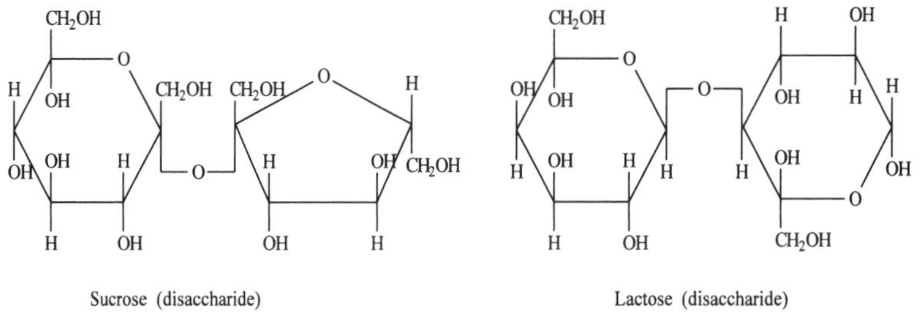

Sucrose (disaccharide)   Lactose (disaccharide)

Figure 7.3
Complex sugars

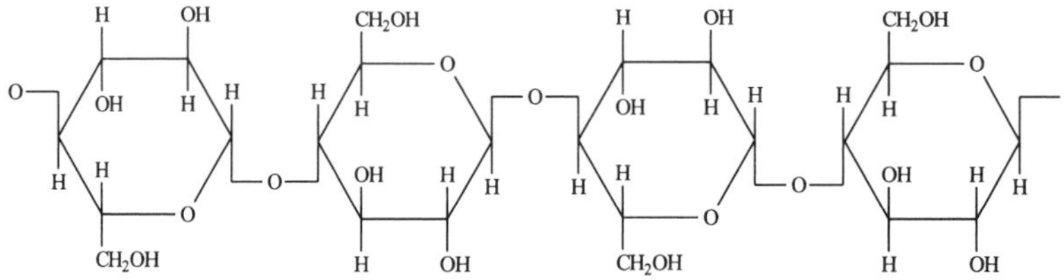

Figure 7.4
Section of a
cellulose
molecule
(polysaccharide)

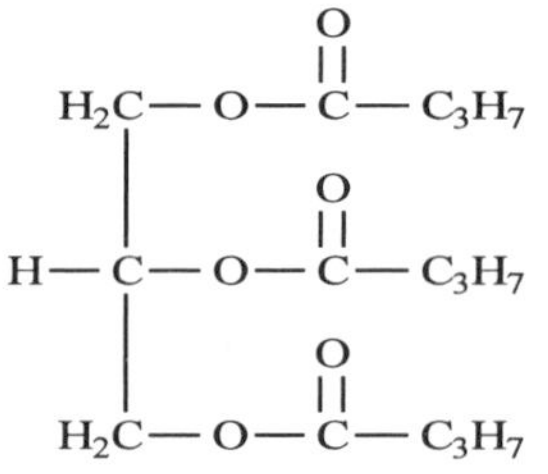

$$H_2C - O - \overset{\overset{\displaystyle O}{\|}}{C} - C_3H_7$$

$$H - \overset{|}{C} - O - \overset{\overset{\displaystyle O}{\|}}{C} - C_3H_7$$

$$H_2C - O - \overset{\overset{\displaystyle O}{\|}}{C} - C_3H_7$$

Glycerol tributyrate

Figure 7.5
Glycerol
tributyrate

Fats comprise a group of organic substances that have in common the property of being soluble to varying extents in organic solvents, while being only sparingly soluble in water. Because of their limited solubility, biodegradation occurs at a slow rate. Fats and oils are both glycerides of fatty acids. Those that are liquid at ambient temperatures are oils while those that are solid are called fats. Figure 7.5 shows a typical molecular structure.

Proteins are complex compounds of carbon, hydrogen, oxygen and nitrogen. Phosphorus and sulphur are present in a few. They are widely distributed in plants and animals. Amino acids constitute the basic building block of proteins. Since there are about 27 known amino acids, the variety of proteins is considerable. Figure 7.6 shows part of a protein molecule containing a string of four amino acids.

7.3.1 Biodegradability of organic substances

The majority of carbohydrates, fats and proteins found in wastewaters are biodegradable. The first step in the biodegradation process is one of

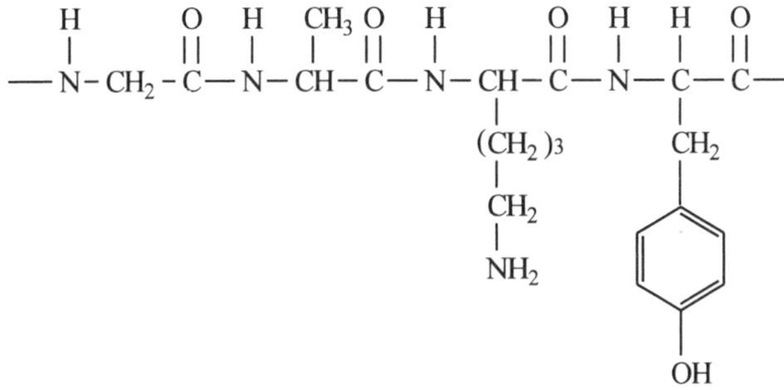

$$\underset{\substack{| \\ (CH_2)_3 \\ | \\ CH_2 \\ | \\ NH_2}}{\underset{}{}} \quad -N-CH_2-\overset{\overset{O}{||}}{C}-\overset{\overset{H}{|}}{N}-\overset{\overset{CH_3}{|}}{CH}-\overset{\overset{O}{||}}{C}-\overset{\overset{H}{|}}{N}-CH-\overset{\overset{O}{||}}{C}-\overset{\overset{H}{|}}{N}-\overset{\overset{H}{|}}{C}-\overset{\overset{O}{||}}{C}-$$

Figure 7.6
Section of a
protein molecule

hydrolysis in which carbohydrates are converted to simple sugars, fats to short-chain fatty acids, and proteins to amino acids.

Several organic compounds, e.g. cellulose, long-chain saturated hydrocarbons, and other complex compounds such as the various synthetic plastics used in industry, are effectively non-biodegradable in the biological wastewater treatment context. Both in the context of their removal in biological wastewater treatment and in relation to their influence on the environment to which they may be discharged, it is useful to classify dissolved organics on a biodegradability scale. Clearly such a classification would range from the easily degradable, e.g. food processing wastes, to the biologically inert, e.g. plastics (toxic wastes are considered later in section 7.4), with a whole range of intermediate degrees of degradability. It would therefore seem desirable to have a continuous scale of measurement on which biodegradability could be quantified. One such scale is that proposed by Thompson *et al.* (1969), whose biochemical treatability index (*BTI*) is calculated as follows:

$$BTI = t_{O5} + t_{Or} + t_{O40} + t_{C50} + t_D$$

where

t_{O5} = time for 5% of the theoretical oxygen demand of the organic chemical to be satisfied;

t_{Or} = time for the theoretical oxygen demand of the organic chemical to be satisfied at the maximum measured rate of oxygen utilization;

t_{O40} = time measured for 40% of the theoretical oxygen demand of the organic chemical to be satisfied;

t_{C50} = time for 50% carbon removal to be reached; and

t_D = time of enzyme activity.

Time measurements are in hours and are related to controlled experimental procedures. Treatability indices for a number of organic substances examined in this way are given in Table 7.2.

While a numerical index, such as that presented in Table 7.2, is a useful and convenient guide to relative biodegradability, it should be noted that biochemical metabolism is a complex process. In particular, it is subject to variation within and between species and is influenced by the effects of acclimation and environment. Experience with biological action on specific wastes shows that acclimation and biodegradation are more likely to occur in a full-scale treatment process than in a bench-scale unit, which in turn is likely to show more activity than a BOD test incubation.

Table 7.2 Biochemical treatability index (*BTI*) for some organic chemicals

Group name	Chemical	Formula	BTl number*
Alcohols	Ethanol	CH_3CH_2OH	14
	Methanol	CH_3OH	240
	n-Propanol	$CH_3CH_2CH_2OH$	12
	Isopropanol	$CH_3CH(OH)CH_3$	117
	Ethylene glycol	$HOCH_2CH_2OH$	43
	tert-Butanol	$(CH_3)_3COH$	321
Acids	Benzoic acid	C_6H_5COOH	21
	Maleic acid	$HOOCCH : CHCOOH$	219
	Oxalic acid	$HOOCOOH$	400
Aldehydes	Benzaldehyde	C_6H_5CHO	32
Esters	Ethyl acetate	$CH_3COOCH_2CH_3$	38
Ethers	Diethylene glycol	$(HOCH_2CH_2)_2O$	180
Ketones	Acetone	CH_3COCH_3	265
	Methyl ethyl ketone	$CH_3COCH_2CH_3$	174
Organic Nitrogens	Acrylonitrile	$CH_2 : CHCN$	483
	Aniline	$C_6H_5NH_2$	232
	Monoetlanolamine	$H_2NCH_2CH_2OH$	320
	Triethanolamine	$(CH_2CH_2OH)_3N$	123
	Pyridine	C_5H_5N	194
Phenols	*o*-Cresol	$CH_3C_6H_4OH$	90
	Hydroquinone	$C_6H_4(OH)_2$	184

* increasing index number indicates more resistance to biodegradation.
Source: Thompson *et al.* (1969)

In general, high molecular weight materials and tertiary-branched molecular structures are not susceptible to metabolism at a significant rate (Ludzack and Ettinger, 1960). Acclimation may have little effect where structures do not permit enzyme approach or fail to diffuse through cell membranes. Atoms in a chain other than carbon, such as oxygen, sulphur and nitrogen, frequently decrease availability.

The COD: BOD$_5$ ratio is a useful simple index of biodegradability, subject to the foregoing comments on the potential limitations related to a lack of acclimation. The COD: BOD$_5$ ratio for untreated municipal sewage is typically in the range 2–3, while that for a biologically treated municipal sewage effluent may be in the range 4–5.

7.3.2 Surfactants

One group of synthetic organics, namely surfactants/detergents or surface-active agents, merit special mention because of their widespread use as aqueous cleaning agents. Although the designation detergent is often used interchangeably with surfactant, it more accurately refers to products that combine surfactants with others substances, organic or inorganic, formulated to enhance cleaning performance. Because of their extensive and increasing use in the home and in industry, surfactants occur in significant concentrations in sewage and industrial wastewaters.

Surfactants are characterized by the following distinctive features (Kirk-Othmer, 1985):

Amphipathic molecular structure—typically containing an oil-soluble hydrocarbon chain and a water-soluble ionic or polar group.

Solubility—a surfactant is soluble in at least one phase of a liquid system.

Adsorption at interfaces—the concentration of surfactant at a phase interface is greater than its concentration in the bulk solution at equilibrium.

Orientation at interfaces—surfactant molecules and ions form oriented monolayers at phase interfaces.

Micelle formation—surfactants form aggregates of molecules or ions called micelles when the concentration of the surfactant in the bulk of the solution exceeds a limiting value

Functional properties—surfactant solutions exhibit combinations of cleaning, foaming, wetting, emulsifying, solubilizing, and dispersing properties.

The presence of two structurally dissimilar groups—a lyophobic (solvent hating) group and a lyophilic (solvent liking) group—within the same molecule is the most fundamental characteristic of surfactants. The terms polar and non-polar are also used to designate water-soluble and water-insoluble groups, respectively.

Surfactants are classified into anionic, cationic, non-ionic and amphoteric categories. In anionic surfactants the hydrocarbon component carries a negative charge, while in cationic surfactants it carries a positive charge. In non-ionic surfactants there is no charge on the molecule, while in amphoteric surfactants the molecule contains both positive and negative charges. The following are typical examples of molecular composition:

anionic: $\quad\quad\quad\quad\quad$ $C_{17}H_{35}CO_2^- Na^+$

cationic: $\quad\quad\quad\quad\quad$ $(C_{18}H_{37})_2^+ N(CH_3)_2 Cl^-$

non-ionic: $C_{15}H_{31}(OC_2H_4)_7OH$

amphoteric: $C_{12}H_{25}^+N(CH_3)_2CH_2CO_2^-$

The hydrophobic part of the molecule typically contains a chain of 10–20 carbon atoms and may include oxygen atoms, amides, esters and other functional groups. The hydrophilic part of anionic surfactants include carboxylates, sulfonates, sulphates and phosphates. Cationics are solubilized by the amine and ammonium groups. Ethylene oxide chains and hydroxyl groups are the solubilizing agents in non-ionic surfactants. Amphoteric surfactants are solubilized by combinations of anionic and cationic solubilizing groups. The molecular weight of surfactants ranges from a low of *ca.* 200 to a high in the thousands for polymeric structures.

Environmental concerns relating to surfactants are focused mainly in two areas: (a) biodegradability and (b) the potential contribution of their phosphorus content to water eutrophication.

The biodegradability of the surfactant molecule is related to the structure of the hydrocarbon radicle. Straight chain hydrophobes (LAS—linear alkanesulphonate) are readily biodegraded (soft), while highly branched hydrocarbon chains (ABS—alkylbenzenesulfonate) are resistant to biodegradation (hard). The cationic detergents, which are salts of quaternary ammonium hydroxide, are not widely used. They are bactericidal and hence are used for both disinfection and cleansing. Non-ionic detergents are derived from polymers of ethylene oxide and are typically resistant to biodegradation.

Biochemical processes (trickling filter or activated sludge process) are generally used to separate non-toxic dissolved organics.

7.4 DANGEROUS SUBSTANCES

The designation dangerous, in this context, implies a capacity to damage living organisms by virtue of toxicity, persistence or bioaccumulation. Environmental regulations are generally framed in a manner that prohibits the discharge of such substances into the aquatic environment. For example, the European Union Directive (76/464/EEC) on pollution caused by certain dangerous substances discharged into the aquatic environment of the Community lists such substances in two categories. List 1 contains those substances which should not be discharged into the aquatic environment, while List 2 contains substances the discharge of which into the aquatic environment should be rigidly controlled.

List 1 substances include:

(1) organohalogen compounds and substances which may form such compounds in the aquatic environment;

(2) organophosphorus compounds;

(3) organotin compounds;

(4) substances in respect of which it has been proved that they possess carcinogenic properties in or via the aquatic environment;

(5) mercury and its compounds;

(6) cadmium and its compounds;

(7) persistent mineral oils and hydrocarbons of petroleum origin;

(8) persistent synthetic substances which may float, remain in suspension or sink and which may interfere with any use of the waters.

List 2 substances include:

(1) metalloids and metals and their compounds, including: zinc, copper, nickel, chromium, lead, selenium, arsenic, antimony, molybdenum, titanium, tin, barium, beryllium, boron, uranium, vanadium, cobalt, thalium, tellurium, silver;

(2) biocides and their derivatives not appearing in List 1;

(3) substances that have a deleterious effect on the taste and/or smell of the products for human consumption derived from the aquatic environment, and compounds liable to give rise to such substances in water;

(4) toxic or persistent organic compounds of silicon, and substances that may give rise to such compounds in water, excluding those that are biologically harmless or are rapidly converted in water into harmless substances;

(5) inorganic compounds of phosphorus and elemental phosphorus;

(6) non-persistent mineral oils and hydrocarbons of petroleum origin;

(7) cyanides, fluorides;

(8) substances that have an adverse effect on the oxygen balance, particularly ammonia and nitrites.

REFERENCES

APHA, AWWA, WEF (1992) *Standard Methods for the Examination of Water and Wastewater*, 18th edn, Washington.

European Union (1976) *Council Directive on Pollution Caused by Certain Dangerous Substances Discharged Into the Aquatic Environment of the Community*, (76/464/EEC), Brussels.

Ludzack, F. J. and Ettinger, M. B. (1960) WPCF, **32**, No. 11, 1173–1200.

Kirk-Othmer, (1985) *Concise Encylcopaedia of Chemical Technology*, John Wiley & Sons Inc., New York.

Sawyer, C. N. and McCarty, P. L. (1967) *Chemistry for Sanitary Engineers*, McGraw-Hill Book Co., New York.

Thompson, C. H., Ryckman, D. W. and Buzzell, J. C. (1969) *Proc. 24th. Ind. Waste. Conf.*, Purdue University., Pt 1, p. 413

8

Adsorption

8.1 INTRODUCTION

Physical adsorption is a process in which solute molecules (adsorbate) become attached to a solid surface under the attracting influence of surface forces (van der Waals force). Thus it is primarily a surface phenomenon. Good adsorbents have a very high specific surface area, which is relatively free of adsorbed materials (it is said to be 'active' or 'activated'). Many organic materials found in water and wastewaters can be removed by adsorption including detergents which have a particularly high surface affinity (surface-active agents). Hydrophilic substances and ions are not amenable to removal by adsorption.

The adsorbent of choice in water and wastewater treatment is activated carbon, which may be used in a dispersed powder form (PAC) or in a fixed-bed granular form (GAC). Its main uses are the removal of taste and odour-causing trace organics and toxic trace organic residues from drinking water and as a final polishing process in advanced wastewater treatment systems.

8.2 ACTIVATED CARBON

Activated carbons suitable for water process applications are produced (Sontheimer *et al.*, 1988) from a variety of raw materials, including bituminous coal, peat, lignite, petrol coke, wood and coconut shells. The production process involves the pyrolytic carbonization of the raw material during which the volatile components are released and the carbon realigns to form a pore structure that is developed during the activation process. The activation process selectively removes carbon, resulting in an opening of closed pores and an increasing in the size of micropores. Activation may be carried out by chemical or physical processes. Chemical activation is normally used for raw materials that contain cellulose and combines both carbonization and activation. It involves pyrolytical heating in the presence of dehydrating chemicals such as zinc chloride or phosphoric acid. Physical activation is more commonly used for the carbons produced for water treatment applications. It involves contact of the carbonized char with steam (carbon dioxide or air are sometimes used) at a temperature in the range 850–1000°C. A maximum surface area per unit mass of original char is found at an activation burnoff in the range 40–50%.

Some of the key physical properties of activated carbon particles are summarized in Table 8.1. The solid fraction density of the carbon matrix is approximately equal to the density of graphite. The particle density is much lower than the matrix density owing to the space occupied by the pore volume. The particle porosity is the ratio of the pore volume to the total volume. The pores vary in size and may be classified (Gregg and Sing, 1982) in size ranges as follows:

Table 8.1 Physical properties of AC particles

Raw material	Solid fraction density (kg m^{-3})	Particle density (kg m^{-3})	Particle porosity
Bituminous coal	1945	491	0.75
Peat/coal	1864	561	0.70
Peat	1981	626	0.68
Coke	1767	783	0.45

Source: Sontheimer *et al.* (1988).

$$\text{micropores: } r_p < 1 \text{ nm}$$
$$\text{mesopores: } 1 \text{ nm} < r_p < 25 \text{ nm}$$
$$\text{macropores: } 25 \text{ nm} < r_p$$

where r_p is the pore radius.

Perhaps the most important physical characteristic of activated carbons is the very large specific surface area associated with their porous structure. (range 600–1500 m^2 g^{-1}). It is important to bear in mind that most of this active surface is contained within the pores of the material, over 99.9% in the case of GAC and obviously less in the case of PAC.

PACs are finely ground, resulting in very small particles with correspondingly large surface areas. The AWWA (1978) standard for the maximum particle size distribution for PAC is as follows:

$$99\% < 149 \text{ μm}$$
$$95\% < 74 \text{ μm}$$
$$90\% < 44 \text{ μm}$$

The grain size distribution for GAC applications may be specified in the same manner as filter sands (see section 6.8), using effective size (d_{10}) and uniformity coefficient (d_{60}/d_{10}) parameters. The effective GAC grain size may be in the range 0.6–1.2 mm and the uniformity coefficient should be less than 2.1 (AWWA, 1974).

8.3 ADSORPTION PROCESS

In the adsorption process there is a mass transfer of solute from a solvent on to the surface of a solid adsorbant. The driving force may be the lyophobic (i.e. solvent-rejecting) character of the solute or the affinity of the solute for the solid, or a combination of both. Thus, the substances that are removed from waters and wastewaters by adsorption on to activated carbon are typically organic contaminants which may have a molecular structure comprised of both hydrophobic and hydrophilic parts. In such cases the hydrophobic part tends to be active at the surface and undergoes adsorption. Surfactants are the prime examples of this category of substance. The second driving force element referred to above , namely the affinity of the solute for the solid, may be due to a surface charge effect, a mass attraction effect or a chemical reaction effect.

The rate at which adsorption proceeds is considered to be diffusion-controlled. Thus, while adsorption equilibrium is quickly attained on exposed particle surfaces, the rate of equilibrium attainment on pore walls is slower, being governed by the rate of diffusion of adsorbate molecules through the capillary pore passages. Since virtually all the useful adsorptive surface area is within the pores the overall rate of adsorption is dependent on particle size—it varies reciprocally with the square of the particle diameter, increases with increasing concentration of solute, decreases with increase in temperature and decreases with decreasing molecular weight of solute (Eckenfelder, 1966). The rate of adsorption has also been found to be proportional to the square root of the time of contact (Morris and Weber, 1964). In addition it is affected by pH, decreasing with an increase in pH and being very poor at pH values above 9.0 (Culp and Culp, 1971).

8.4 ADSORPTION EQUILIBRIUM RELATIONSHIPS

The mathematical formulation of adsorption relationships is usually expressed in terms of either Freundlich or Langmuir equations.

The Freundlich equation has the form:

$$\frac{X}{M} = kC^{1/n} \tag{8.1}$$

where X is the mass of the adsorbate, M is the mass of the adsorbant, C is the equilibrium concentration of the substance remaining in solution, k and n are constants. The equation is more useful in logarithmic form :

$$\log \frac{X}{M} = \log k + 1/n \log C$$

A plot of X/M versus C on log–log paper results in a straight line of slope $1/n$. Such a graph is called an adsorption isotherm—it relates, for a particular temperature, the mass of substance adsorbed per unit mass of adsorbent with the concentration of substance in the effluent. An adsorption isotherm can be determined experimentally in the laboratory, using powdered activated carbon, which is separated out by

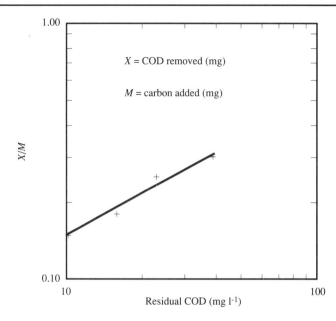

X = COD removed (mg)

M = carbon added (mg)

X/M

0.10
10   100

Residual COD (mg l⁻¹)

Figure 8.1
Example of
adsorption
isotherm

filtration after a suitable contact period. A typical test result is shown in Figure 8.1. Such tests are a useful indication of the potential of the adsorption process for a particular application, but design data are preferably obtained from pilot plant operation.

The Langmuir equilibrium adsorption equation has the form:

$$\frac{X}{M} = \frac{abC}{(1 + aC)} \tag{8.3}$$

which may also be written as

$$\frac{1}{X/M} = \frac{1}{b} + \frac{1}{ab} \cdot \frac{1}{C} \tag{8.4}$$

where b is the amount of adsorbate required to form a complete monolayer on the adsorbent surface and is a constant that increases with molecular size. While a and b values have been reported for a number of organic substances, the usefulness of the Langmuir equation in water and wastewater applications of adsorption is limited since many organics are usually adsorbed simultaneously.

8.5   DESIGN OF ADSORPTION SYSTEMS

8.5.1   PAC process design

While GAC has replaced PAC in many continuous process applications, PAC processes offer considerable design flexibility and economy in situations where an adsorption step is required only on an intermittent basis, as may be the case in many drinking water treatment applications.

Powder activated carbon (PAC) may be used in a single-stage or multi-stage application mode. Because of the small particle size, adsorption equilibrium is reached rapidly. If it is assumed that adsorption equilibrium is reached in each stage of a multi-stage process the specific removal (X/M) in each stage will, according to the Freundlich equation (8.1), be proportional to the $(1/n)$th power of the residual concentration ($c^{1/n}$). Thus, there will be a step reduction in the X/M value for each succeeding stage, the lowest value being in the final stage. If, however, only a single stage is used, its X/M value must be the same as that of the final stage of the multi-stage process if it is to achieve the same final residual concentration as the multi-stage process. Hence, the average specific removal rate (average X/M) will be much higher for a multi-stage process than for a single stage process. Despite the process advantage of using more than a single stage, most practical applications are single-stage processes.

PAC is applied at many waterworks on an occasional basis for taste and odour control, typically at a dose in the range 5–10 g m^{-3}. If more than 25 g m^{-3} has to be added for long periods of time, GAC columns are likely to be more economical. It is typically added at the rapid mix stage of chemical coagulation processes and removed in the downstream sedimentation and sand filtration processes. This mode of application provides the essential process components of particle wetting, adsorption contact time and particle separation. It should be noted that PAC use in this manner is likely to increase filter head loss and reduce filter run time. PAC may also be separated by membrane filtration processes.

PAC has been applied in wastewater treatment in conjunction with the activated sludge process (Schultz and Keinath, 1984), where it is added to the influent in the aeration basin at concentrations in the range 50–300 g m^{-3}. Its observed positive effects (Sontheimer *et al.*, 1988) include improved process stability, increased mixed liquor suspended solids concentration (i.e. increased microbial biomass) and improvements in the settling and dewatering characteristics of the activated sludge. In the case of wastewaters containing toxic components, some of the enhanced removal efficiency can be attributed to the adsorption of the toxic compounds by the PAC.

8.5.2 GAC process design

Granular activated carbon (GAC) in fixed beds is generally preferred to its use in powdered form, where continuous application is required. GAC allows a more complete use of the adsorption capacity of the carbon, thus reducing make-up costs. GAC columns provide a filtration capacity as well as an adsorption function. GAC is easier to handle than

PAC, requiring only to be replaced when its adsorption capacity is reached, typically after three months to one year of operation.

GAC columns may be designed to operate in a conventional down-flow filtration mode or in an upflow mode. The residence time in GAC columns is usually expressed in terms of the empty bed contact time or EBCT, which is generally in the range 5–20 min. Most economic use of granular carbon can be made in upflow columns operated on the countercurrent principle (Culp and Culp, 1971). Spent carbon is removed from the bottom of the bed periodically and a corresponding quantity of fresh carbon is added to the top of the bed. With this system, optimum removal is effected and maximum use is made of the adsorp-tive capacity of the carbon at the same time. Packed-bed and expanded-bed systems are used. Packed beds can only be used for waters of low turbidity (less than 2.5 JU) owing to their susceptibility to clogging. The upflow velocity of turbid waters in expanded-bed systems must be sufficient to provide a self-cleansing action. For a grain size of 0.8–1.0 mm this can be achieved at a velocity of about 15 m h^{-1} with a corresponding bed expansion of about 10% (Hager and Flentje, 1965). Upflow operation has the advantage that the spent carbon is always at the bottom where it can be easily removed. This remains true in expanded-bed operation owing to the fact that the effective density of the grains increases with the quanitity of adsorbate attached and hence, owing to its increased density, there is a tendency for saturated material to migrate towards the bottom of the bed.

A batch-operated fixed-bed system, in which all the carbon is replaced when breakthrough occurs, may also be used. Saturation of the carbon moves progressively down through the bed until finally breakthrough occurs as shown in Figure 8.2. When the carbon has to serve as filter and adsorbent, the bed is cleaned by backwashing as in ordinary rapid sand filtration. However, back-washing of a downflow system tends to upset bed stratification and the advantages of counter-current operation are lost.

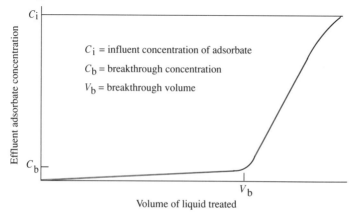

C_i = influent concentration of adsorbate

C_b = breakthrough concentration

V_b = breakthrough volume

Volume of liquid treated

Figure 8.2
Illustration of carbon breakthrough

Two important considerations should be borne in mind in relation to the design of adsorption systems: (a) adsorption is a process for removing dissolved solids, and for the best results the influent turbidity should be as low as possible; and (b) reliable values for design parameters—volume of carbon required to secure the desired effluent quality, rate of carbon saturation—can only be reliably obtained through pilot plant investigation.

Spent or saturated carbon can be reactivated by a number of methods, including solvent washing, acid or caustic washing, steam treatment and thermal regeneration. The latter method is the most widely used.

REFERENCES

American Waterworks Association (AWWA) (1974), *AWWA Standard for Granular Activated Carbon, 604–74*

American Waterworks Association (AWWA) (1978), *AWWA Standard for Powdered Activated Carbon*, 600–78

Culp, R. L. and Culp, G. L. (1971) *Advanced Wastewater Treatment*, Van Nostrand Reinhold Co., New York.

Eckenfelder, W. W. (1966), *Industrial Water Pollution Control*, McGraw-Hill Book Co., New York.

Gregg, S. J. and Sing, K. S. W. (1982) *Adsorption, Surface Area and Porosity*, Academic Press, London.

Hager, D. G. and Flentje, M. E. (1965), *J. Am Waterworks Assoc.* **57**, 1440

Morris, J. C. and Weber, W. J. (1964), *Adsorption of Biochemically Resistant Materials from Solution*, USPHS AWTR–9.

Schultz, J. R. and Keinath, T. M. (1984) *J. Wat. Pollut. Control. Fed.*, **56**, 143.

Sontheimer, H. Crittenden, J. C. and Summers, R. (1988) *Activated Carbon for Water Treatment*, DVGW- Forschungsstelle, Engler-Bunte-Institut, Universitt Karlsruhe, Germany.

Additional reading

McGuire, M. J. and Suffet, I. H., eds. (1983) *Treatment of Water by Granular Activated Carbon*, Advances in Chemistry Series 202, American Chemical Society, Washington, DC.

9

Chemical Precipitation

9.1 INTRODUCTION

The main chemical processes used in water and wastewater treatment practice include

(1) Chemical coagulation, which is used to remove colloidal material.

(2) Chemical precipitation, which is used in water treatment to reduce water hardness by precipitation of calcium and magnesium ions and, in wastewater treatment, to remove phosphate as a precipitate of calcium, magnesium or iron.

(3) Oxidation/reduction reactions, which, in conjunction with precipitation, are used to remove metals from wastewaters.

(4) Water stabilization, which regulates water pH to bring its carbonate system to an equilibrium state.

(5) Ion-exchange processes, which are used for water softening and demineralisation.

Chemical coagulation has been discussed in Chapter 3, while ion-exchange is the subject of Chapter 10. The remaining chemical processes are discussed in this chapter.

The following physical elements are common to most chemical process systems:

(a) chemicals storage and handling facilities;

(b) chemicals dosing system;

(c) rapid mixing system;

(d) reaction vessel (generally agitated); and

(e) solids separation system.

Storage and handling facilities depend to a large extent on (i) the form and quantity in which chemicals are purchased, i.e. whether in bulk, liquid, solid or powdered form, and (ii) the characteristics of the chemicals, i.e. whether corrosive, hygroscopic etc. It is important to note that the most widely used construction materials—steel and concrete—require special protection in a low pH environment.

Chemical processes may be operated on a batch or on a continuous basis. For continuous operation special dosing facilities are required. Dry feeders may be used for the continuous addition of non-hygroscopic chemicals in powdered form, while variable-stroke piston or diaphragm pumps are widely used for the continuous addition of solutions. Simple dosing facilities are adequate for batch process operations.

The mixing system must effect a uniform dispersion of the added chemical in the water or wastewater being treated. Mixing systems which may be used include conduit mixing (conduit velocity > 1 m s^{-1}) ; pump mixing,—where the chemical is injected into the suction side of the pump; and turbine, propellor or other impeller-type, high-speed mixing devices.

The size of the reaction vessel required depends on reaction kinetics. In general, process design should aim to optimize the reaction kinetics so that the desired reaction goes to completion in the minimum time. This often requires that the reaction pH has to be maintained at a fixed value by the addition of acid (e.g. coagulation of protein, isoelectric pH 4.2) or alkali (e.g. precipitation of calcium and magnesium). Where precise control of the pH is required the use of automated control systems is essential.

Chemical precipitates are separated by sedimentation, sometimes with the aid of prior coagulation.

9.2   PRECIPITATION PROCESSES

The extent of separation of a cation by precipitation depends to a large extent on the solubility of its precipitate, which is usually either a hydroxide, carbonate or phosphate salt. At the saturation or equilibrium concentration of these slightly soluble salts an equilibrium exists between the solute ions and the solid phase, e.g.

$$CaCO_3(s) \Leftrightarrow Ca^{2+} + CO_3^{2-} \qquad (9.1)$$

for which the equilibrium expression is

$$\frac{[Ca^{2+}][CO_3^{2-}]}{[CaCO_3]_s} = K \qquad (9.2)$$

The activity of the solid phase is constant and hence the denominator in equation (9.2) can be combined with K:

$$[Ca^{2+}][CO_3^{2-}] = K_{sp} \qquad (9.3)$$

where K_{sp} is called the solubility product. Solubility products for a number of slightly soluble salts relevant to the present discussion are given in Table 9.1. In the case of hydroxides, solubilities can be calculated directly from the appropriate K_{sp} value if the pH is known. For carbonates and phosphates the calculation of solubility is complicated by their reactions in solution:

Table 9.1 Solubility Constants at 25°C*

Solid	pK$_{sp}$	Solid	pK$_{sp}$
Fe(OH)$_3$ (amorph)	38	BaSO$_4$	10
FePO$_4$	17.9	Cu(OH)$_2$	19.3
Fe$_3$(PO4)$_2$	33	PbCl$_2$	4.8
Fe(OH)$_2$	14.5	Pb(OH)$_2$	14.3
FeS	17.3	PbSO$_4$	7.8
Fe$_2$S$_3$	88	PbS	27
Al(OH)$_3$ (amorph)	33	MgNH$_4$PO$_4$	12.6
AlPO$_4$	21	MgCO$_3$	5
CaCO$_3$ (calcite)	8.34	Mg(OH)$_2$	10.7
CaCO$_3$ (aragonite)	8.22	Mn(OH)$_2$	12.8
CaMg(CO$_3$)$_2$ (dolomite)	16.7	AgCl	10
CaF$_2$	10.3	Ag$_2$CrO$_4$	11.6
Ca(OH)$_2$	5.3	Ag$_2$SO$_4$	4.8
Ca$_3$(PO$_4$)$_2$	26	Zn(OH)$_2$	17.2
CaSO$_4$	4.59	ZnS	21.5
SiO$_2$ (amorph)	2.7		

* The equilibrium constant for the reaction $A_z B_{y(s)} \Leftrightarrow zA^{y+} + yB^{z-}$ is $K_{sp} = [A^{y+}]^z [B^{z-}]^y$; $pK_{sp} = -\log K_{sp}$.
Source: Snoeyink and Jenkins (1980)

$$CO_3^{2-} + H_2O \Leftrightarrow HCO_3^- + OH^- \tag{9.4}$$

$$PO_4^{3-} + H_2O \Leftrightarrow HPO_4^{2-} + OH^- \tag{9.5}$$

for which the the equilibrium expressions are usually written in the form:

$$\frac{[H^+][CO_3^{2-}]}{[HCO_3^-]} = K_2 \tag{9.6}$$

$$\frac{[H^+][PO_4^{3-}]}{[HPO_4^{2-}]} = K_3 \tag{9.7}$$

Taking CaCO$_3$ as an example:

$$[Ca^{2+}] = [CO_3^{2-}] + [HCO_3^-] \tag{9.8}$$

also

$$[H^+][OH^-] = K_w \tag{9.9}$$

The simultaneous solution of equations (9.3), (9.6), (9.8) and (9.9), together with the approximation that $[OH^-] = [HCO_3^-]$, produces the following equation:

$$K_2 S^4 - 2K_2 K_{sp} S^2 - K_w K_{sp} S + K_2 K_{sp}^2 = 0 \qquad (9.10)$$

where $S = [Ca^{2+}]$

For a phosphate salt the corresponding equation is:

$$K_3 S^4 - 2K_3 K_{sp} S^2 - K_w K_{sp} S + K_3 K_{sp}^2 = 0 \qquad (9.11)$$

9.2.1 Precipitation of Ca²⁺ and Mg²⁺

Ca^{2+} and Mg^{2+} are mainly responsible for causing 'hardness' in water. They precipitate soaps, hindering lather formation and may also cause the formation of a hard scale in hot water distribution systems and boilers. Water hardness is generally expressed as equivalent $CaCO_3$ in mg l^{-1}. Waters may be classified in hardness terms as follows:

0–75 mg l^{-1}	soft
75–150 mg l^{-1}	moderately hard
150–300 mg l^{-1}	hard
> 300 mg l^{-1}	very hard

Ca^{2+} and Mg^{2+} can be precipitated in carbonate, hydroxide or phosphate forms. Carbonates and hydroxides are produced in the lime and lime–soda processes of water softening. On addition of hydrated lime ($Ca(OH)_2$) to a hard water the following reactions occur:

$$Ca(OH)_2 + Ca(HCO_3)_2 \rightarrow 2CaCO_3 \downarrow + 2H_2O \qquad (9.12)$$

$$Ca(OH)_2 + Mg(HCO_3)_2 \rightarrow CaCO_3 \downarrow + MgCO_3 \downarrow + 2H_2O \qquad (9.13)$$

The respective solubilities of $CaCO_3$ and $MgCO_3$ at 20°C are about 8.4 and 110 mg l^{-1}, respectively. In the presence of excess lime, $MgCO_3$ is converted to $Mg(OH)_2$ which has a solubility of about 8.4 mg l^{-1} at 20°C:

$$MgCO_3 + Ca(OH)_2 \rightarrow CaCO_3 \downarrow + Mg(OH)_2 \downarrow \qquad (9.14)$$

When Ca^{2+} and Mg^{2+} are associated with anions other than carbonates (non-carbonate hardness), precipitation can be effected by the addition of sodium carbonate:

$$Na_2CO_3 + CaSO_4 \rightarrow Na_2SO_4 + CaCO_3 \downarrow \qquad (9.15)$$

$$Na_2CO_3 + MgSO_4 + Ca(OH)_2 \rightarrow Mg(OH)_2 \downarrow + CaCO_3 \downarrow + Na_2SO_4 \qquad (9.16)$$

It is worthy of note that in the removal of hardness by the lime process the ion content of the water is reduced, while in the removal of non-carbonate hardness in the lime-soda process Ca^{2+} and Mg^{2+} are simply exchanged for Na^+.

In theory, it should be feasible to reduce the hardness of water by chemical precipitation to a level corresponding to the solubilities of the carbonate and hydroxide precipitates produced, i.e. to a level of about 30 mg l^{-1} as $CaCO_3$. In practice, however, the limiting concentration attainable is about 50–70 mg l^{-1} as $CaCO_3$. This may be due to the formation of complex magnesium carbonates, which are more soluble than the hydroxide form.

Water softened by the lime or lime–soda processes are not stable and hence require pH adjustment, as discussed in section 9.3.

Sodium phosphate is also used as a precipitating agent for Ca^{2+} and Mg^{2+} ions. The resulting calcium and magnesium phosphates are less soluble than the corresponding carbonates and hydroxides (< 3.0 mg l^{-1} as $CaCO_3$). The following are the relevant reactions:

$$3Ca(HCO_3)_2 + 2Na_3PO_4 \rightarrow Ca_3(PO_4)_2 + 6NaHCO_3 \qquad (9.17)$$

$$3CaSO_4 + 2Na_3PO_4 \rightarrow Ca_3(PO_4)_2 + 3Na_2SO_4 \qquad (9.18)$$

Because phosphate is a relatively expensive precipitant, it can sometimes be used to best economic advantage in combination with lime and sodium carbonate when a water of very low hardness is required.

9.2.2 Precipitation of phosphorus from wastewaters

Phosphorus and nitrogen are the key nutrients that cause enrichment of natural water bodies giving rise to excessive growth of aquatic plants and plant-like organisms such as algae, a condition generally known as eutrophication. In most fresh waters, phosphorus is the nutrient that limits growth because of its limited availability in the natural aquatic environment. Hence, the reduction of phosphorus to a low level is normally specified for wastewaters discharged into such sensitive receiving waters. This can be achieved by chemical precipitation or by enhanced biological uptake (see chapter 12 for discussion of latter).

The typical total phosphorus (TP) concentration in municipal wastewater varies from conurbation to conurbation but generally falls within the range 6–15 mg l^{-1}. Conventional wastewater treatment, including primary sedimentation and normal biological treatment, is unlikely to remove more than about 2 mg l^{-1}. Hence, in circumstances where there is a requirement for the reduction of wastewater TP to a low level, say 1–2 mg l^{-1} (discharge to sensitive waters), it is necessary to include a phosphorus precipitation step in the treatment system. The chemical precipitants that are used include hydrated lime, aluminium and iron salts. The equilibrium solubilities for the resultant phosphorus precipitates are plotted in Figure 9.1

The precipitant chemical may be added to the raw influent (pre-precipitation) or may be dosed into the mixed liquor in the aeration basin of an activated sludge process (simultaneous precipitation) or in a separate final stage of treatment (post-precipitation).

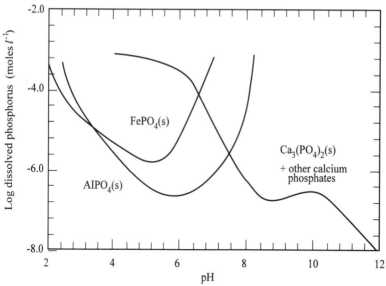

Figure 9.1
Equilibrium
solubility plots for
iron, aluminium
and calcium
phosphates.
(Reprinted with
permission from
Jenkins and
Hermanourèz
1991. © Lewis
Publishers, an
imprint of CRC
Press, Boca Raton,
Florida)

Precipitation as calcium phosphate is achieved by adding lime $(Ca(OH)_2)$ to the wastewater, raising the pH to about 10.5. The lime dose required to elevate the pH to this level is a function of the alkalinity and may be approximated as 1.5 times the alkalinity as $CaCO_3$. Thus, the calcium dose is not stoichiometrically related to the phosphorus concentration. While lime is a relatively low-cost chemical, the disadvantages of the lime process are that it produces a large amount of sludge residue and creates too high a pH for environmental discharge. While it would appear from the solubility characteristics of calcium phosphates, plotted in Figure 9.1, that a low phosphate concentration could be achieved at a pH value in the range 8.5–9.5, this has not been found to be feasible under typical wastewater process operating conditions.

Iron and aluminium salts are used for phosphorus-removal in both simultaneous and post-precipitation processes. Operating experiences (Casey *et al.*, 1978; USEPA, 1987) show that the minimum solubility levels, plotted in Figure 9.1, are not achievable in wastewater treatment precipitation processes at metal ion dose rates corresponding to the stoichiometric metal ion:phosphate ion ratio. It is found in practice that a metal ion addition of two to three times the stoichiometric requirement is necessary to reduce the effluent TP to the region of 1–2 mg l^{-1}. This excess over the stoichiometric amount is due (Jenkins and Hermanowicz, 1991) to the simultaneous precipitation of metal hydroxide and the formation of soluble metal phosphate complexes.

9.2.3   Reduction and precipitation of Cr^{6+}

Hexavalent chromium occurs in wastes from metal plating and anodizing processes. The pH of such wastes is usually quite low and the Cr^{6+}

concentration may be as high as 2% by weight. Removal is effected in a two-stage process, in which Cr^{6+} is first reduced to Cr^{3+} and then precipitated as a hydroxide. The reducing agents commonly used are sulphur dioxide (SO_2), ferrous sulphate ($FeSO_4$) and sodium metabisulphite ($Na_2S_2O_5$). The following oxidation–reduction reactions occur:

$$\text{oxidation:} \qquad 3SO_2 + 6H_2O \rightarrow 3SO_4^{2-} + 12H^+ + 6e^- \qquad (9.19)$$

$$\text{reduction:} \qquad Cr_2O_7^{2-} + 14H^+ + 6e^- \rightarrow 2Cr^{3+} + 7H_2O \qquad (9.20)$$

$$\text{oxidation/reduction } Cr_2O_7^{2-} + 3SO_2 + 2H^+ \rightarrow 2Cr^{3+} + 3SO_4^{2-} + H_2O \qquad (9.21)$$

Reaction (9.21) is highly pH-dependent, being virtually instantaneous below pH 2.0. The reaction with metabisulphite is essentially the same as with SO_2. When ferrous sulphate is the reducing agent used, the following reactions occur:

$$\text{oxidation:} \qquad 6Fe^{2+} \rightarrow 6Fe^{3+} + 6e^- \qquad (9.22)$$

$$\text{reduction:} \qquad Cr_2O_7^{2-} + 14H^+ + 6e^- \rightarrow 2Cr^{3+} + 7H_2O \qquad (9.23)$$

$$\begin{array}{l} \text{oxidation/} \\ \quad \text{reduction:} \qquad 6Fe^{2+} + Cr_2O_7^{2-} + 14H^+ \rightarrow 6Fe^{3+} + 2Cr^{3+} + 7H_2O \end{array} \qquad (9.24)$$

Reaction (9.24) is rapid below pH 3.0

The trivalent chromium produced on reduction of Cr^{6+} is precipitated as a hydroxide on the addition of hydrated lime:

$$Cr_2(SO_4)_3 + 3Ca(OH)_2 \rightarrow 2Cr(OH)_3 \downarrow + 3CaSO_4 \qquad (9.25)$$

The ferric iron from reaction (9.24) is similarly precipitated:

$$Fe_2(SO_4)_3 + 3Ca(OH)_2 \rightarrow 2Fe(OH)_3 \downarrow + 3CaSO_4 \qquad (9.26)$$

9.3 CHEMICAL STABILIZATION

In the context of water treatment, stability is generally taken to mean the existence of an equilibrium condition in respect of dissolved calcium carbonate, i.e. a stable water will neither deposit calcium in the distribution system nor will it take calcium carbonate into solution from the distribution system. When such a condition exists, equilibrium equations (9.3) and (9.6) apply:

$$[Ca^{2+}][CO_3^{2-}] = K_{sp} \qquad (9.3)$$

$$\frac{[H^+][CO_3^{2-}]}{[HCO_3^-]} = K_2 \qquad (9.6)$$

Combining (9.3) and (9.6):

$$\frac{[Ca^{2+}][HCO_3^-]}{[H^+]} = K_{sp}/K_2 \qquad (9.27)$$

or

$$pH_s = pK_2 - pK_{sp} - \log [Ca^{2+}] - \log [HCO_3^-] \qquad (9.28)$$

where pH_s is the equilibrium pH corresponding to the Ca^{2+} and HCO_3^- concentrations in solution. The difference beween the pHs of a water and its actual pH is a measure of its instability and is sometimes called the Langelier saturation index (I):

$$I = pH - pH_s$$

In a fully stable water, I is zero. When I is positive the water is supersaturated and will deposit $CaCO_3$; when I is negative, the water is undersaturated and will tend to take $CaCO_3$ into solution.

If the Ca^{2+} and HCO_3^- values are known, the value of pH_s can be calculated from equation (9.30), using the pK_{sp} and pK_2 values given in Table 9.2.

Table 9.2 Stability constants

Temp. (°C)	5	10	15	20	25	40
pK_{sp}	8.09	8.15	8.22	8.28	8.34	8.51
pK_2	10.56	10.49	10.43	10.38	10.33	10.22

The influence of ion concentration on ion activity can be taken into account by modifying the constant values given in Table 9.2 as follows (Fair *et al.*, 1968):

$$pK_{sp} = pK_{sp}(T) - 4(\mu)^{0.5}/(1 + 3.9(\mu)^{0.5}) \qquad (9.29)$$

$$pK_2 = pK_2(T) - 2(\mu)^{0.5}/[1 + 1.4(\mu)^{0.5}] \qquad (9.30)$$

where T refers to temperature and is calculated as follows: $\mu \cong 2.5 \times 10^{-5} S_d$ where S_d is the total dissolved solids in the water (mg l^{-1}), or

Figure 9.2 Pipe samples from water distribution systems, showing scale formation in samples (a) and (b) and modular corrosion in sample (c). (Reproduced courtesy of Dr. P. O'Connor, Department of Civil Engineering, University College, Dublin).

$\mu \cong 4H - T_{alk}$, where H is the total hardness (moles l^{-1}) and T_{alk} is the total alkalinity (eq l^{-1}).

Some natural waters of low hardness and alkalinity as well as coagulated waters softened by the ion-exchange process may have a negative saturation index. This is corrected by the addition of a hydroxide, usually calcium hydroxide. Lime-softened waters, on the other-hand, have a positive saturation index and are equilibrated by the addition of an acid or CO_2 gas.

REFERENCES

Casey, T. J., O'Connor, P. E. and Greene, R. (1978), *Trans. Instit. of Eng. Ireland*, **103**, 13–20.

Fair, G. M., Geyer, J. C. and Okun, D. A. (1968) *Water and Wastewater Engineering*, **2**, John Wiley & Sons, Inc., New York.

Jenkins, D. and Hermanowicz, S. W. (1991) Principles of chemical phosphate removal, in *Phosphorus and Nitrogen Removal from Municipal Wastewater*, ed. Sedlak, R., Lewis Publishers, Boca Raton, Florida.

Snoeyink, V. L. and Jenkins, D. (1980) *Water Chemistry*, John Wiley & Sons, Inc., New York.

USEPA (1987) *Handbook: Retrofitting POTWs for Phosphorus Removal in the Chesapeake Bay Drainage Basin*, EPA/625/6- 87/017.

Related reading

Eilbach, W. J. and Mattock, G. (1987) *Chemical Processes in Wastewater Treatment*, Ellis Horwood Ltd, Chichester, UK.

Sawyer, C. N. and McCarty, P. L. (1967) *Chemistry for Sanitary Engineers*, 2nd. edn., McGraw-Hill Book Co., New York.

Stumm, W. and Morgan, J. J. (1981) *Aquatic Chemistry*, John Wiley & Sons Inc., New York.

10

Ion Exchange

10.1 INTRODUCTION

Ion exchange is the reversible interchange of ions between a solid ion-exchange medium and a solution. Cations may be exchanged for Na^+ or H^+ while anions are exchanged for OH^-. Most exchange media in current use are insoluble synthetic polymer resins, although some naturally occurring minerals such as the greensand zeolites (analcite, clinoptilolite, montmorillonite and others) are also used. Synthetic media possess a very large number of charged functional groups, such as $-SO_3^-$ and $-NH_3^+$, to which are loosely attached (by electrostatic forces) small mobile ions of opposite charge. These latter ions are exchanged for ions of the same sign (so-called counter-ions) in solution. Ions of the same charge as the functional groups are known as co-ions.

Resins are designated as cation exchangers or anion exchangers, depending on the counter-ions they exchange. Some resins can exchange both cations and anions and are termed amphoteric ion exchangers.

Depending on bead structure and size, resins may be designated as gel resins, macroporous resins, isoporous resins and micro resins. Gel resins and macroporous resin beads are typically 0.3–1.2 mm in diameter and have a three-dimensional matrix of macromolecular hydrocarbon chains, usually consisting of a copolymer of styrene and divinylbenzene. Gel resins have a homogeneous non-porous structure. They imbibe water, which usually results in a swelling of the resin. Ions diffuse through the imbibed water and are exchanged for the mobile ions attached to the polystyrene chains. The matrix of the porous resins is of the same structure as that of the gel resins but the existence of pores allows a more ready access of water into the beads and thus reduces the resistance to diffusion and enables a higher rate of ion exchange. Micro resins are resins in powder form, used in a somewhat similar manner to powdered activated carbon and also requiring separation by a filtration process. The properties of a typical ion-exchange resin are given in Table 10.1.

10.2 EXCHANGE EQUILIBRIUM

The selectivity and extent of exchange depend on the equilibrium relationships between ions in solution and those attached to the solid phase.

Within a sodium cation exchanger, through which a solution containing Ca^{2+} ions passes, the following exchange takes place:

Table 10.1 Properties of a typical strong acid resin

		Gel structure	Macroporous structure
Resin bead diameter	(mm)	0.3–1.2	0.3–1.2
Bulk density	(kg m^{-3})	849.5	833.5
Moisture content	(%)	45–48	40–46
pH range		0–14	0–14
Maximum temperature	(°C)	120	150
Turbidity tolerance	(NTU)	5	5
Chlorine tolerance	(mg l^{-1})	1	1
Back-wash rate	(m^3 h^{-1} m^{-2})	12.2	14.7
Back-wash period	(min)	20	20
Service rate	(m^3 h^{-1} m^{-3})	16–50	16–50
Regeneration rate	(m^3 h^{-1} m^{-3})	4	4
Rinse volume	(m^3 m^{-3})	3	3–4
Total capacity	(keq m^{-3})	1.5	1.8

Source: Bolto & Pawlowski (1987)

$$2(Na^+R^-) + Ca^{2+} \Leftrightarrow (Ca^{2+}R_2^{2-}) + 2Na^+ \qquad (10.1)$$

where R represents the negatively charged polymer. When the available exchange sites are effectively used up, the medium can be regenerated by passing a concentrated solution of Na^+ ions through the bed (5–10% NaCl):

$$(Ca^{2+}R_2^{2-}) + 2Na^+ \Leftrightarrow 2(Na^+R^-) + Ca^{2+} \qquad (10.2)$$

Similarly, for a hydrogen cation exchange medium:

$$2(H^+R^-) + Ca^{2+} \Leftrightarrow (Ca^{2+}R^{2-}) + 2H^+ \qquad (10.3)$$

Regeneration with H_2SO_4 (2–10% solution):

$$(Ca^{2+}R^{2-}) + 2H^+ \Leftrightarrow 2(H^+R^-) + Ca^{2+} \qquad (10.4)$$

An ion exchange medium replaces anions in solution by OH^- ions:

$$SO_4^{2-} + 2(R^+OH^-) \Leftrightarrow 2OH^- + (R^{2+}SO_4^{2-}) \qquad (10.5)$$

Regeneration with NaOH (5–10% solution):

$$(R^{2+}SO_4^{2-}) + 2OH^- \Leftrightarrow 2(R^+OH^-) + SO_4^{2-} \qquad (10.6)$$

The selectivity of an exchange medium for ions in solution can be quantified in equilibrium terms. The equilibrium expression corresponding to exchange equation (10.1) is

$$\frac{[Ca^{2+}R_2^{2-}][Na^+]^2}{[Na^+R^-]^2[Ca^{2+}]} = K_s(NaR \rightarrow CaR) \qquad (10.7)$$

where K_s is a selectivity coefficient, defining the relative distribution of counter-ions when the cation exchanger in the Na^+ form is placed in a solution containing the Ca^{2+} cations. K_s typically has a value close to unity for the exchange of monovalent ions by monovalent ions and between 20 and 40 for the exchange of monovalent ions by bivalent ions, thus indicating, with some exceptions, a greater affinity for bivalent ions than for monovalent ions (Fair *et al.*, 1968). It should be noted, however, that K_s is not a constant for all resins, being dependent on a number of factors, such as the degree of cross-linking in the resin, the valency of the ions and the concentration of the solution (Helfferich, 1962).

10.3 SELECTIVITY AND CAPACITY OF EXCHANGE RESINS

Information on the qualitative order of affinity for counter-ions is essential for practical design purposes. The following are typical examples (Bolto and Pawlowski, 1987):

sulphonic acid resin: $Fe^{3+} > Al^{3+} > Ca^{2+}$
carboxylic acid resin: $H^+ > Ca^{2+} > Mg^{2+} > K^+ > Na^+$
quaternary ammonium resin: $NO^{3-} > CrO_4^{2-} > Br^- > Cl^-$
polyamine resin:
$$OH^- > SO_4^{2-} > CrO_4^{2-} > NO_3^- > PO_4^{3-} HCO_3^- \geqslant Br^- > Cl^- > F^-$$

It follows from equation (10.7) that the extent of the selective removal of Ca^{2+} ions from a solution containing Ca^{2+} and Na^+ ions depends on the relative concentrations of Ca^{2+} and Na^+ and increases greatly as the solution becomes more dilute. This is illustrated graphically in Figure 10.1.

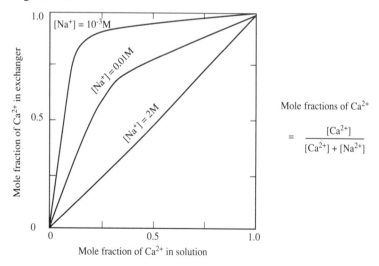

Mole fractions of Ca^{2+}

$$= \frac{[Ca^{2+}]}{[Ca^{2+}] + [Na^{2+}]}$$

Figure 10.1
Influence of concentration on selective ion removal (Reprinted by permission of John Wiley & Sons, Inc. From Foir *et al.*, 1968)

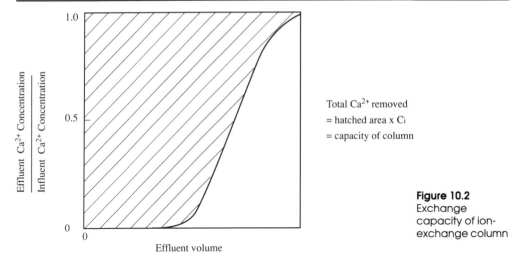

Figure 10.2
Exchange
capacity of ion-
exchange column

The capacity of an ion-exchange medium is measured by the number of charges it can replace per unit volume of medium, the unit of measurement being eq l^{-1}. The exchange capacity of a given material can be determined experimentally by measuring the total quantity of ions exchanged, when the exchange reaction is driven to completion in a column containing a known volume of the medium. A typical result of such an experiment is shown on Figure 10.2. The capacity of synthetic media is generally within the range 0.5–2.0 eq l^{-1}.

10.4   ION EXCHANGE APPLICATIONS

In the water treatment field ion exchange is used for water softening and demineralization. A common system in use is the down-flow fixed bed arrangement, using a granular medium, as illustrated in Figure 10.3. The typical properties of such a resin are given in Table 10.1.

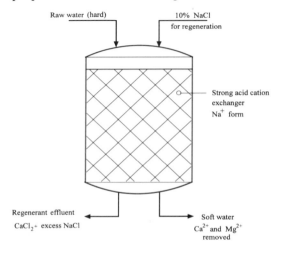

Figure 10.3
Schematic
arrangement of
ion-exchange
column for water
softening

In softening plants a strong acid sodium cation exchange medium is normally used, while in demineralizing plants, beds of H^+ and OH^- media may be used in series or in a single mixed bed. To avoid short-circuiting, beds are commonly not less than 0.8 m in depth. The required bed volume for a particular application can be most reliably obtained from the results of a laboratory column study. Flow rates obtaining in practice generally fall within the range 0.2–1.0 m^3 min^{-1} m^{-3} of medium. When breakthrough occurs, the bed is generally back-washed with clean water to remove any accumulated solids before regeneration is carried out. Regeneration and back-wash rates are generally as given in Table 10.1.

The influent to an ion exchanger must be low in turbidity and in organics which might be adsorbed by the exchange medium, thus inactivating its exchange capability.

REFERENCES

Bolto, B. A. and Pawlowski, L. (1987) *Wastewater Treatment by Ion Exchange*, E & F Spon, London.

Fair , G. M., Geyer, J. C. and Okun, D. A. (1968) *Water and Wastewater Engineering*, **2**, John Wiley & Sons Inc., New York.

Helfferich, F. (1962) *Ion Exchange*, McGraw-Hill Book Co., New York.

11

Biological Processes

11.1 INTRODUCTION

The primary objective of biological wastewater treatment processes is the conversion of biodedgradable organics into a microbial biomass which can be separated by appropriate solids/liquid separation processes such as sedimentation, flotation, etc..

Most organic wastewaters contain relatively low concentrations of organic matter and can be dealt with efficiently and economically by aerobic treatment processes, in which part of the organic matter is converted to carbon dioxide through microbial respiration and part is converted to microbial biomass residue.

More concentrated wastewaters and organic suspensions, such as sewage sludge, can also be effectively stabilized anaerobically. Anaerobic wastewater treatment converts organic matter to methane and carbon dioxide and also to an anaerobic biomass residue.

The technology of biological processes is concerned with the design of reactor vessels, which provide an optimum environment for microbial growth and in which a high active microbial biomass concentration can be developed. Aerobic process units require a continuous input of oxygen to support microbial respiration, while oxygen must be completely excluded from anaerobic wastewater treatment processes, being acutely toxic to the methanogenic bacteria.

Biological wastewater treatment processes may be of the suspended floc type, the so-called activated sludge processes, or of the attached film type, the so-called biofilter systems. In suspended floc systems (discussed in detail in Chapter 12), the active microbial biomass forms a dispersed aqueous suspension, with which the waste stream is brought into contact by a mixing system. In attached film processes (discussed in detail in Chapter 13), the active biomass is attached as a film to a solid medium of stone or plastic material. The wastewater is brought into contact with the active biomass while flowing over the medium surface as a thin stream. Aerobic filters are typically operated in a trickling downflow mode, which permits free air movement within the filter medium. They can also be operated in submerged downflow mode with a counter-current air flow to provide the necessary oxygen input to meet the microbial respiration demand. Anaerobic filters are generally operated in an upflow flooded mode, effectively excluding air.

11.2   FACTORS AFFECTING MICROBIAL GROWTH

11.2.1   Energy and cell synthesis

Microbial growth (Gaudy and Gaudy, 1980) results from the conversion of dissolved organic matter plus certain inorganic trace elements into cell protoplasm through a complex train of metabolic reactions. The terms respiration and fermentation are commonly applied to those metabolic reactions that produce the energy required for cell synthesis.

All living organisms use a common form of energy storage. Whatever the energy source, energy derived from that source is stored as chemical bond energy in the form of adenosine triphosphate (ATP). The hydrolysis of ATP to adenosine diphosphate (ADP) is an exergonic (energy-producing) reaction, which releases about 7000 calories per mole of ATP. Organisms couple the energy released upon hydrolysis of ATP with the endergonic (energy-requiring) reactions associated with the synthesis of cellular macromolecules such as proteins, lipids, polysaccharides and nucleic acids. This is illustrated in Figure 11.1, which shows the synthesis of compound AB through the coupling of endergonic reaction $A + B \rightarrow AB$ with the exergonic reaction $ATP \rightarrow ADP + P_i$, where P_i denotes inorganic phosphate.

The energy required to form ATP from ADP is derived from the biochemical degradation of organic compounds. In respiring organisms this energy is released through substrate oxidation, the electrons released being passed via electron-carriers to a terminal electron acceptor. Under aerobic conditions the terminal electron acceptor is oxygen ($O_2 \rightarrow H_2O$), while in non-aerobic respiration the terminal electron acceptor may be nitrate ($NO_3^- \rightarrow N$) or sulphate ($SO_4^{2-} \rightarrow S^{2-}$).

The generation of ATP through fermentation takes place under anaerobic conditions and involves the production of an energy-rich phosphorylated intermediate which is capable of donating its phosphate group to ADP to form ATP. Fermentation is an inefficient

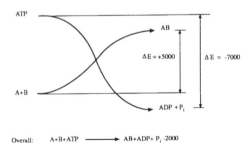

Overall: $A + B + ATP \longrightarrow AB + ADP + P_i$ -2000

Figure 11.1
Coupling of synthesis with hydrolysis of ATP

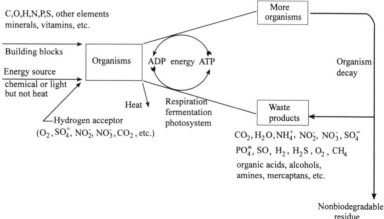

C,O,H,N,P,S, other elements
minerals, vitamins, etc.

Figure 11.2
Schematic
representation of
biological processes

process for releasing the available free energy in the substrate, leaving most of it in the fermentation products. For example, in the fermentation of glucose to form lactic acid only about 2% of the potentially available energy in glucose is captured for ATP formation. ATP formation is thus the major growth-limiting factor in fermentation systems.

The large difference in energy production efficiency between respiration and fermentation systems has important consequences in biological wastewater treatment systems. Aerobic systems, served by efficient energy-releasing respiration, generate large amounts of active biomass residue in the form of surplus sludge, while anaerobic processes, deriving their energy by the fermentation route, generate far less biomass and produce energy-rich methane gas as an end-product.

A schematic representation of biological processes is shown on Figure 11.2.

11.2.2 Nutrient requirements

Water is the major constituent of microbial cells (75–90% by weight). The elemental composition of the cellular solid fraction varies somewhat, depending on environmental conditions and the species of microorganism. The typical cell composition for the widely distributed bacterium *Escherichia coli* is presented in Table 11.1.

The four elements, carbon, oxygen, nitrogen and hydrogen, make up more than 90% of the cell dry weight. These elements, plus phosphorus and sulphur, constitute the macromolecules of the cell. The remaining 4% of the cell dry matter includes a large number of elements — potassium, sodium, calcium, magnesium, chlorine, iron, manganese, cobalt, copper, boron, zinc, molybdenum and others. Porges (1956) suggested the formula $C_5H_7NO_2$ for the stoichiometric elemental proportions of a heterogeneous microbial population.

Table 11.1 Typical elemental cell composition for Escherichia coli

Element	Dry weight (%)
Carbon	50
Oxygen	20
Nitrogen	14
Hydrogen	8
Phosphorus	3
Sulphur	1
Potassium	1
Sodium	1
Calcium	0.5
Magnesium	0.5
Chlorine	0.5
Iron	0.2
All others	0.3

Source: Stanier *et al.* (1976)

Heterotrophic microorganisms can utilize a great variety of organic compounds as sources of cell carbon, while autotrophic organisms use carbon dioxide as the sole carbon source. The cell nitrogen requirements may be derived from organic or inorganic nitrogen sources. Inorganic nitrogen forms include ammonia (NH_3), nitrate (NO_3^-), nitrite (NO_2^-) and gaseous nitrogen (N_2). Ammonia is the most readily utilizable form of inorganic nitrogen. A limited number of bacteria are capable of using gaseous nitrogen as the nitrogen source (the nitrogen-fixing bacteria). A general classification of microorganisms by their sources of energy and carbon is given in Table 11.2.

Table 11.2 Classification of microorganisms by energy and carbon source

Designation	Energy source	Carbon source
Autotrophic:		
photosynthetic	light	CO_2
chemosynthetic	inorganic oxidation–reduction reaction	CO_2
Heterotrophic	organic oxidation–reduction reaction	Organic carbon

The requirements for the major nutrients nitrogen and phosphorus are generally considered to be satisfied (Pipes, 1979; Speece and McCarty, 1964) by the following threshold nutrient ratios:

	BOD$_5$: N : P
aerobic processes:	100 : 5 : 1
anaerobic processes:	100 : 0.5 : 0.1

11.2.3 Influence of temperature

Of the physical factors affecting microbial growth in any environment, temperature is one of the most influential in the selection of species. Temperature affects growth in two opposing ways:

(1) According to the Arrhenius relationship (equation (1.5)), an increase in temperature speeds up chemical enzymatic reactions and, therefore, microbial growth and product formation;

(2) With rising temperature, proteins, nucleic acids and other cellular components that are sensitive to temperature, i.e. temperature labile, will tend to become irreversibly deactivated and lysis, death and endogenous metabolism rates will increase. Hydrolysis rates will also increase with temperature.

For every organism there is:

(1) a minimum temperature below which no growth occurs;

(2) an optimum temperature at which the organism grows most rapidly; and

(3) a maximum temperature above which no growth occurs.

The temperature range over which microbial growth can occur is usually considered to be between $-12°C$ and $+120°C$ (Hamer, 1995; Sonnleitner and Fiechter, 1983). While no single microbe can grow throughout this range, most organisms are eurythermal, i.e. they are capable of growing over a fairly wide temperature range of some 30–40°C. The characteristic relationship between the specific growth rate constant μ and temperature is shown in Figure 11.3 (the specific growth rate $\mu(d^{-1})$ is correlated with doubling time t_d according to the relation: $\mu = \ln 2/t_d$). This plot is characterized by four growth zones: (a) zone I extends from the lowest temperature for growth to the temperature at which growth rate increases at a logarithmic rate; (b) zone II is the temperature range of logarithmic growth rate increase, as predicted by

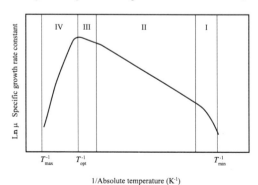

Figure 11.3
Influence of
temperature on
microbial growth
rate

the Arrhenius relationship for chemical reactions (equation (1.5)); (c) zone III is the optimum temperature range, which extends from the upper temperature limit for logarithmic growth to the temperature of maximum growth rate; and (d) zone IV is the super-optimum range, in which the growth rate declines rapidly with increasing temperature. This latter range typically does not extend more than about 6°C above the temperature of maximum growth rate. For most organisms the growth rate increases two- to threefold for each 10°C rise in temperature in the logarithmic temperature range (zone II in Figure 11.3).

   The main groups of organisms, classified on the basis of temperature, are:

(1) Psychrophiles (temperaure optima < 10°C).

(2) Psychrotrophs or facultative psychrophiles (grow well below 10°C, but have temperature optima > 10°C).

(3) Mesophiles (temperature optima 10°C–40°C).

(4) Thermotolerant strains (temperature optima 45°C–60°C).

(5) Thermophiles (temperature optima > 60°C).

(6) Caldoactive strains (temperature optima > 75°C).

(7) Barothermotolerant strains (temperature optima < 100°C, maximum > 100°C).

(8) Barothermophiles (temperature optima > 100°C, maximum > 100°C)

11.2.4   Influence of pH

Each microbial species can grow within a specified pH range, which typically extends over 3 or 4 pH units, with optimum growth rate near the midpoint of the range.

   Most bacteria grow in the pH range 5–9 and have a pH optimum near to neutrality.

   Most fungi prefer an acid environment with a pH optimum near 5.

   The anaerobic degradation of carbohydrates, producing organic acids, causes a lowering of the environmental pH to the region 4.5–5.0, as a result of which fermentation is inhibited or stopped. In such cases fermentation acts as a preservative process as, for example, in the ensilage of grass to provide winter feed for cattle.

   The sulphur-oxidizing bacteria (Thiobacillus), which oxidize reduced sulphur compounds to sulphate (SO_4^{2-}), creating a highly acidic environment (production of H_2SO_4), are capable of growth at pH 1.0 or less.

   The anaerobic methane-producing bacteria, which have a particular importance in wastewater treatment, are very sensitive to pH. Their pH growth range is 6.4–7.6, which is unusually narrow.

11.2.5 Oxygen and microbial growth

Organisms that require dissolved oxygen for survival are classified as obligate aerobes. Those that cannot grow in the presence of oxygen are classified as obligate anaerobes. Those that can grow with or without oxygen are classified as facultative anaerobes.

Obligate aerobes and facultative anaerobes are to be found among the bacteria, fungi and protozoa. Bacteria constitute the main group of obligate anaerobes, although some protozoa are also obligate anaerobes.

Dissolved oxygen can be toxic to aerobic microorganisms when present at supersaturated concentration. Anaerobic organisms are usually quite sensitive to oxygen although a few are aerotolerant. Obligate anaerobes are killed by exposure to oxygen. This sensitivity to oxygen may be extreme, as in the case of the methane bacteria.

11.3 KINETICS OF MICROBIAL GROWTH

A knowledge of kinetics, i.e. the rate of conversion of biochemical substrate to end-products and the factors that determine these rates, is essential as a rational basis for process design. The kinetics of microbial growth in the mixed culture environment, characteristic of biological wastewater treatment reactors, are normally based on an adaptation (Lawrence and McCarty, 1970) of the kinetic model of Monod (1950), which was derived for continuous-growth pure-culture systems.

Consider a suspended microbial biomass growing in a liquid medium containing the necessary nutrients for growth. The rate of change in microbial concentration $X (\text{mg l}^{-1})$ with time $t(\text{d})$ is the net sum of the growth and decay processes:

net microbial growth rate
$(\text{mg l}^{-1} \text{ d}^{-1})$:
$$\frac{dX}{dt} = \mu X - k_d X \qquad (11.1)$$

where μ is the specific growth rate coefficient (d^{-1}) and k_d is the decay coefficient (d^{-1}). According to the Monod growth model, the dependence of the specific growth rate coefficient on substrate concentration is expressed as follows:

$$\mu = \hat{\mu} \left(\frac{S}{K_s + S} \right) \qquad (11.2)$$

where $\hat{\mu}$ is the maximum specific growth rate constant at the saturation concentration of growth-limiting substrate (d^{-1}), S is the substrate concentration (mg l^{-1}), and K_s is the saturation constant, which is the substrate concentration at which the specific growth rate is half the maximum specific growth rate. Combining equations (11.1) and (11.2) results in the following expression for the net microbial growth rate:

net microbial growth rate
(mg l^{-1} d^{-1})

$$\frac{dX}{dt} = \frac{\hat{\mu}XS}{K_s + S} - k_d X \qquad (11.3)$$

The rate of reduction of the substrate concentration S is related to the microbial growth rate by the mass balance relation:

$$-\frac{dS}{dt} = \frac{\hat{\mu}XS}{Y(K_s + S)} \qquad (11.4)$$

where Y is the yield coefficient.

The dependence of the specific growth rate μ on substrate concentration S (equation (11.2)) is illustrated graphically in Figure 11.4. The shape of this curve is influenced by the values of the two regulating constants, the saturation constant K_s and the maximum specific growth rate $\hat{\mu}$. In particular, the saturation constant K_s is a key parameter in relation to microbial competition for food at low substrate concentration, such as may be the case in wastewater treatment processes producing well-stabilized effluents. In such growth-limiting environments, those organisms with the greatest affinity for the substrate (low K_s values, i.e. rapid growth at low substrate concentration) will outgrow those with the least affinity for the substrate (high K_s values), to become the predominant species in the microbial population. At high substrate concentrations, on the other hand, the growth rate is largely determined by the magnitude of the maximum specific growth rate constant, $\hat{\mu}$. These characteristic effects of K_s and $\hat{\mu}$ are illustrated in Figure 11.4, which shows the specific growth rate variations for two heterotrophic organisms, A and B, as a function of substrate concentration. Organism A has a lower K_s value than organism B and hence grows more rapidly than organism B at low substrate concentration. Organism B has a

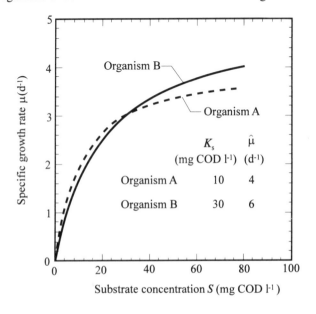

Figure 11.4
Influence of substrate concentration on specific microbial growth rate

higher $\hat{\mu}$ value than organism A and hence grows more rapidly than organism A at high substrate concentration.

It is important to emphasize that the foregoing mathematical model of the microbial growth process is a gross simplification of what is acknowledged to be a very complex process. The complexity is compounded in biological wastewater treatment processes due to the heterogeneous nature of both substrate and microbial population. The situation is further complicated by the fact that the model parameters can be expressed in a variety of units. For example, the organic carbon substrate concentration may be expressed in a number of alternative units, including BOD, COD, TOC and volatile solids. Substrate nitrogen compounds are conventionally expressed in equivalent nitrogen concentration units, e.g. NH_3–N, NO_3–N. Because it is not feasible to measure the concentration of living organisms in the microbial biomass, the value assigned to the parameter X is usually the volatile fraction of the suspended biomass concentration. It is not surprising, therefore, that the values reported in the literature for the foregoing kinetic coefficients show a wide variation, depending on process environmental conditions and wastewater composition.

Typical model parameter values for the activated sludge process (heterotrophic and autotrophic growth) are given in Table 11.3, while a corresponding set of typical parameter values for the anaerobic digestion process (methanogenic organism growth) are given in Table 11.4.

Table 11.3 Typical model parameter values for the activated sludge process

Parameter	Unit	Value at 20°C	Value at 10°C
Y_A	g cell COD formed (g N oxidized)$^{-1}$	0.2–0.3	0.2–0.3
Y_H	g cell COD formed (g COD oxidized)$^{-1}$	0.5–0.7	0.5–0.7
$\hat{\mu}_H$	d^{-1}	4–6	2–4
$\hat{\mu}_A$	d^{-1}	0.6–0.8	0.3–0.4
K_S	g COD m^{-3}	10–80	10–80
K_{NO}	g NO_3–N m^{-3}	0.3–0.6	0.3–0.6
K_{NH}	g NH_3–N m^{-3}	0.8–1.2	0.8–1.2
k_d	d^{-1}	0.06–0.1	0.02–0.06

Source: Gray (1990); Horan (1990); IAWPRC (1987).

Table 11.4 Typical model parameter values for the anaerobic digestion process

Parameter	Unit	Value at 25°C	Value at 35°C
$\hat{\mu}_{AN}$	d^{-1}	0.15–0.25	0.30–0.45
K_{AC}	g CH_3COOH m^{-3}*	300–400	100–200
Y	g cell VSS (g CH_3COOH)$^{-1}$	0.04–0.06	0.04–0.06
k_d	d^{-1}	0.01–0.015	0.015–0.02

* Acetic acid.
Source: Metcalf and Eddy Inc. (1974); Bailey and Ollis (1986).

It is clear from the relative values of the kinetic coefficients presented in Table 11.3 that the heterotrophic organisms grow much more rapidly than the autotrophs ($\hat{\mu}_H \gg \hat{\mu}_A$). The low value of $\hat{\mu}_A$ also has implications for the minimum microbial residence time required for autotrophic process applications, such as nitrification, as discussed in section 11.5. It is also noteworthy that the substrate saturation constants for nitrification and denitrification processes, K_{NH} and K_{NO}, respectively, are very low, making these processes virtually zero-order processes with respect to substrate.

The influence of oxygen concentration on aerobic process kinetics may also be modelled by a Monod type function:

$$\frac{dX}{dt} = X(\mu - k_d)\left(\frac{DO}{K_o + DO}\right) \tag{11.5}$$

where DO is the dissolved oxygen concentration (mg l^{-1}) and K_o is the oxygen saturation constant. K_o has a value of about 0.2 mg l^{-1} for heterotrophs and a value of about 0.4 mg l^{-1} for autotrophs.

Much research effort has been devoted to the development of more comprehensive model descriptions of biological growth processes in wastewater treatment. The Task Group on Modelling of Biological Processes of the International Association for Water Pollution Research and Control has developed multi-parameter activated sludge models (IAWPRC, 1987, 1994), incorporating carbon, nitrogen and phosphorus removal. The modelling approach is based on Monod process kinetics and first-order microbial decay, as outlined above, but is extended to take into account the range of conditions encountered in practice. A total of seven dissolved and six particulate components are used to characterize the wastewater and the sludge. In addition to dissolved oxygen and alkalinity, these include two forms of biomass, seven fractions of COD (organic material) and four fractions of nitrogen. Nine transformation processes are included; three relate to the growth of heterotrophic biomass, two represent decay of biomass and four describe 'hydrolysis' processes, in which complex organic matter is made available for biodegradation in the form of simpler molecules (Gujer and Henze, 1991).

11.4 BIOLOGICAL REACTORS

Biological processes are carried out in tank reactors under controlled conditions. Depending on their flow characteristics, reactors can be classified as being of the plug-flow, mixed-flow or arbitrary-flow type. The ideal plug-flow reactor is completely unmixed, all fluid elements having the same retention time—the influent moves through the tank as a plug. The completely mixed reactor is totally devoid of gradients in composition or temperature. The arbitrary-flow reactor has mixing characteristics which fall between complete mix and plug-flow. Many practical reactors fall into this latter category. Tracer output curves, illustrating the flow characteristics of these reactors, are shown on Figure 11.5.

In *plug-flow* reactors with a continuous tracer input the first output trace appears after a time interval of t_d, where t_d is the nominal detention time, i.e. $t_d = V/Q$; the concentration thereafter is the same as the influent concentration C_0. Where the input is a tracer slug, the output is a corresponding slug which emerges after a detention time t_d.

In *completely mixed* reactors with a continuous tracer input, the output profile is found from mass balance considerations, as follows:

$$V\frac{dc}{dt} = Q(C_0 - C) \tag{11.6}$$

which integrates to give:

$$C = C_0(1 - e^{-t/t_d}) \tag{11.7}$$

Where the input to a completely mixed reactor is a tracer slug, the output profile is similarly found from mass balance considerations:

$$-V\frac{dc}{dt} = QC \tag{11.8}$$

which integrates to give:

$$C = C_0 e^{-t/t_d} \tag{11.9}$$

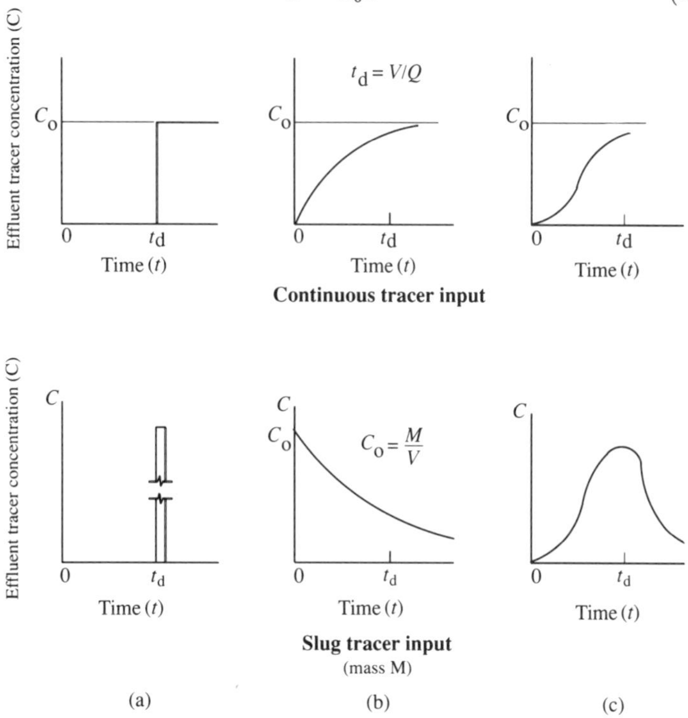

Figure 11.5
Output tracer profiles for step and slug tracer inputs to (a) plug-flow, (b) complete mix, and (c) arbitrary flow reactors (flow rate Q; reactor volume V)

11.5   APPLICATION OF KINETICS TO A MIXED REACTOR

A microbial mass balance equation for a completely mixed reactor of volume V (m^3), operating at steady state, may be written as follows:

$$V\frac{\mathrm{d}X}{\mathrm{d}t} = 0 = VX_t - M_t \tag{11.10}$$

where X_t is the net microbial growth rate and M_t is the rate of removal of microbial biomass from the reactor. Combining equations (11.1), (11.2) and (11.10):

$$M_t = V\left(\frac{\hat{\mu}XS}{K_s + S} - k_\mathrm{d}X\right) \tag{11.11}$$

or

$$\frac{M_t}{VX} = \hat{\mu}\left(\frac{S}{K_s + S}\right) - k_\mathrm{d} \tag{11.12}$$

hence

$$\frac{1}{\theta_s} = \hat{\mu}\left(\frac{S}{K_s + S}\right) - k_\mathrm{d} \tag{11.13}$$

where θ_s is the microbial solids residence time. From equation (11.13) the effluent substrate concentration S may be expressed as a function of θ_S:

$$S = \frac{K_s(1 + k_\mathrm{d}\theta_s)}{\theta_s(\hat{\mu} - k_\mathrm{d}) - 1} \tag{11.14}$$

Thus θ_s, the microbial solids residence time, is identified as a key biological process variable. It will be noted from equation (11.14) that, in a mixed reactor, the substrate concentration S (which is also the effluent substrate concentration) is uniquely determined by θ_s and is independent of the inflow substrate concentration and the hydraulic residence time. S is plotted as a function of θ_s in Figure 11.6 for a mixed aerobic reactor, and in Figure 11.7 for a mixed anaerobic reactor, based on the sets of kinetic and stoichiometric coefficients noted on the diagrams. The plotted results indicate that the microbial residence time required for stable operation in anaerobic processes is an order of magnitude greater than in aerobic processes and also, because of the much higher K_s value for anaerobic processes, it is not feasible to reduce substrate concentration to a very low level by anaerobic processes. As well as the foregoing kinetic differences between aerobic and anaerobic microbial processes, the stoichiometric differences are also of considerable process significance. The relatively high Y value for aerobic processes means that they produce a much higher yield of microbial biomass than anaerobic processes.

It will be noted from Figures 11.6 and 11.7 that the microbial solids residence time converges to a lower limiting value, approaching which there is a steep increase in substrate concentration. This minimum microbial residence time θ_{sm} can be estimated from equation (11.13) by making the assumption that S is large compared with K_S, resulting in the relation:

$$\frac{1}{\theta_{sm}} = \hat{\mu} - k_d \qquad (11.15)$$

On the basis of the kinetic coefficient values for the aerobic and anaerobic systems used in Figures 11.6 and 11.7, respectively, the minimum microbial residence times are calculated to be:

aerobic: $\qquad\qquad\qquad\qquad \theta_{sm} = 0.2$ d
anaerobic: $\qquad\qquad\qquad\qquad \theta_{sm} = 5.6$ d

For efficient process operation, it is necessary to have a high microbial concentration in the reactor. This is commonly achieved by microbial sludge recycle from a downstream sedimentation tank and can also be enhanced by appropriate reactor design so as to maximize microbial biomass retention.

The steady-state microbial biomass concentration X in a mixed reactor of volume V, having a microbial residence time θ_s, an influent volumetric flow rate Q and an influent substrate concentration S_i is derived from the relation:

$$\frac{VX}{YQ(S_i - S) - Vk_d X} = \theta_s \qquad (11.16)$$

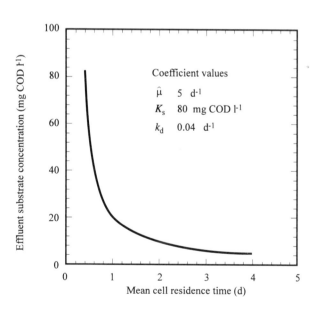

Figure 11.6
Aerobic reactor: influence of mean cell residence time on effluent susbstrate concentration

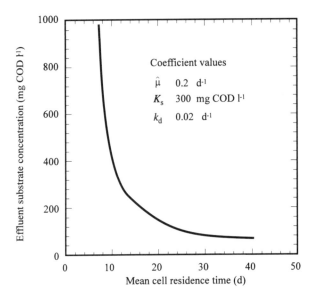

Coefficient values

$\hat{\mu}$ 0.2 d^{-1}

K_s 300 mg COD l^{-1}

k_d 0.02 d^{-1}

Figure 11.7
Anaerobic reactor:
influence of mean
cell residence time
on effluent substrate
concentration

which gives:

$$X = \frac{Y(S_i - S)\theta_s}{\theta_H(1 + k_d\theta_s)} \qquad (11.17)$$

where θ_H is the hydraulic retention time equal to V/Q.

The surplus active microbial biomass (SS) produced under these operating conditions is:

surplus active biomass: $SS = YQ(S_i - S) - Vk_dX \qquad (11.18)$

Combining equations (11.17) and (11.18):

$$SS = \frac{YQ(S_i - S)}{1 + k_d\theta_s} \qquad (11.19)$$

The actual production of excess biological sludge in an activated sludge process will exceed the foregoing estimate of the production of active microbial biomass, owing to the inclusion of inert matter and dead microbial cell residues.

11.6 PLUG-FLOW REACTOR KINETICS

A plug-flow reactor (PFR) differs from a completely mixed reactor (CMR) in that there is a substrate concentration gradient from a maximum value at the inlet end to a minimum value at the outlet end. It follows from this that, where a PFR and a CMR are both producing the same quality effluent under identical operating conditions, the required volume of the PFR will be less than that of the CMR because the average reaction rate in the PFR is higher than the common reaction rate throughout the CMR volume. It is not feasible to quantify the effluent substrate concentration for a PFR in terms of the operational parameters, as has been done for the CMR. However, the retention time (V/Q) required in a PFR in order to reduce the substrate concentration to a specified level can be determined by numerical computation as follows:

$$\Delta V = \frac{Q\Delta S}{dS/dt} \tag{11.20}$$

where ΔV is the reactor volume required to achieve a substrate reduction of ΔS and Q is the flow rate. In a PFR, dS/dt has a maximum value at the inlet end and a minimum value at the outlet end. The hydraulic retention time θ_H may be expressed in the form:

$$\theta_H = \sum \frac{\Delta V}{Q} = \sum \frac{\Delta S}{dS/dt} \tag{11.21}$$

where the summation value for S varies from the influent substrate concentration S_i to the effluent substration concentration S_e. The corresponding expression for a CMR is

$$\theta_H = \frac{S_i - S_e}{dS/dt} \tag{11.22}$$

where the substrate uptake rate is constant and based on the value of S_e.

Figure 11.8 compares the computed relative values of θ_H (and hence also the relative reactor sizes) for PFR and CMR reactors, based on the following assumed values for the kinetic coefficients: $\hat{\mu} = 5.0\ \mathrm{d}^{-1}$; $K_s = 80\ \mathrm{mg\ COD\ 1}^{-1}$; $Y = 0.5$; and $k_d = 0.05\ \mathrm{d}^{-1}$.

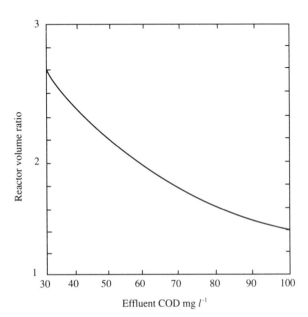

Figure 11.8
CMR:PFR volume ratio to reduce an influent COD concentration from 600 mg l^{-1} to the indicated effluent value

Since the mean specific substrate conversion rate in a CMR reactor is proportional to the effluent substrate concentration, it is to be expected that the volumetric advantage of the PFR reactor over its CMR counterpart increases as the effluent substrate concentration is reduced. As Figure 11.8 indicates, the calculated reactor volume ratio reduces from 2.6 at a COD concentration of 30 mg l^{-1} to a value of 1.4 at a COD value of 100 mg l^{-1}.

REFERENCES

Bailey, J. E. and Ollis, D. F. (1986) *Biochemical Engineering Fundamentals*, 2nd. edn, McGraw-Hill Book Co., New York.

Gaudy, A. and Gaudy, E. (1980) *Microbiology for Environmental Scientists and Engineers*, McGraw-Hill Book Company, New York.

Gray, N. F. (1990) *Biology of Wastewater Treatment*, Oxford University Press, Oxford.

Gujer, W. and Henze, M. (1991) Water. Sci. Technol., **23**, 1011–1023.

Hamer, G. (1995) Some biological aspects of landfills, Workshop on Solid Waste Disposal, University College, Dublin.

IAWPRC Task Group on Mathematical Modelling for Design and Operation of Biological Wastewater Treatment, *Activated Sludge Model No. 1* (1987), *Activated Sludge Model No. 2* (1994), IAWPRC, London.

Horan, N. J. (1990) *Biological Wastewater Treatment Systems*, John Wiley & Sons, Ltd., Chichester, UK.

Lawrence, A. W. and McCarty, P. L. (1970) *J. San. Eng. Div., ASCE*, **96**, No. SA3.

Metcalf and Eddy Inc. (1974) *Wastewater Engineering*, Tata McGraw-Hill Publishing Company Ltd., New Delhi.

Monod, J. (1950) *Ann. Inst. Pasteur*, **79**, 390–410.

Pipes, W. O. (1979) *J. Water Pollat Control Fed.*, **51**, 62–70.

Porges, N., Jasewics, L. and Hoover, S. R. (1956) Principles of biological oxidation, in *Biological Treatment of Sewage and Industrial Wastes*, eds. McCabe and Eckenfelder, Reinhold, New York.

Speece, R. E. and McCarty, P. L. (1964) Nutrient requirements and biological solids accumulation in anaerobic digestion, in *Advances in Water Pollution Research*, **2**, pp. 305–322, Ed. W.W. Eckenfelder, Pergamon Press, New York.

Sonnleitner, B. and Fiechter, A. (1983) *Trends in Biotechnol.*, **1**, No.3, 74–80.

Stanier, R. Y., Adelberg, E. A. and Ingraham, J. (1976) *The Microbial World*, 4th. edn, Prentice-Hall, Englewood Cliffs, N.J.

Related reading

Bitton, G. (1994) *Wastewater Microbiology*, John Wiley & Sons Inc., New York.

WPCF (1990) *Wastewater Biology, the Microlife*, Water Pollution Control Federation, Alexandria, VA, USA.

12

Activated Sludge Processes

12.1 INTRODUCTION

The activated sludge process is generally considered to have had its origins in aeration experiments carried out by Ardern and Lockett in Manchester in 1914 (IWPC, 1987). They found that, when they retained and built up in an aerated vessel the biological floc formed in a series of aeration experiments on sewage, the time required for its purification progressively decreased as the concentration of floc increased. They referred to this floc as being 'activated' and the resulting process became known as the activated sludge process. In the intervening period this process has become one of the main methods, used worldwide for the purification of wastewaters containing biodegradable organic solids, its most important application being the treatment of domestic sewage.

The principal unit in all activated sludge processes is the aeration or biochemical reaction vessel. In it the waste organics are mixed with the active sludge and this so-called 'mixed liquor' is aerated for several hours, during which the microorganisms (mainly bacteria) in the sludge utilise the organic matter in the waste for energy and cell synthesis. After aeration, the activated sludge biomass is separated from the mixed liquor in a secondary settling tank and the separated sludge is recycled back to the aeration tank. A proportion of the settled sludge may be removed from the system on a continuous or intermittent basis to maintain the concentration of active biomass in the aeration unit within the desired value range. The characteristic schematic layout of the activated sludge process is shown in Figure 12.1.

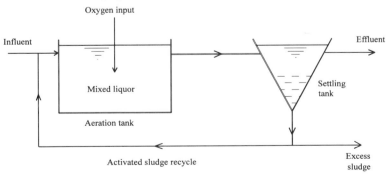

Figure 12.1
Schematic layout of the activated sludge process

The operating mixed liquor suspended solids (MLSS) may vary in the range 1500–5000 mg l^{-1}. The active fraction of the MLSS is conveniently approximated for design purposes as its organic component or mixed liquor volatile suspended solids (MLVSS). The MLVSS:MLSS ratio is typically in the range 0.65–0.85.

As shown in Chapter 11, the microbial biomass residence time or sludge age, θ_S, is a key performance parameter in biological reactors. The θ_S value for an AS process, operating at steady state, is given by the ratio of the sludge mass in the aeration basin to the sludge production rate and is typically expressed in days. A related parameter favoured by designers is the 'sludge loading rate' (L_S), expressed as kg BOD$_5$ per kg MLVSS · d or as kg COD per kg MLVSS · d. Since both L_s and θ_s are used in design computations, it is important to understand the relationship between these two basic parameters. This relationship is developed in the following section, based on substrate measurement as COD (IAWPRC, 1987, 1994).

The yield of volatile suspended solids (kg VSS per kg COD removed) in a completely mixed activated sludge process, operating at a steady state sludge age θ_s, can be considered to be made up of three components:

the influent-derived nonbiodegradable VSS fraction removed: X_o

the active microbial biomass (AMB): $\dfrac{Y \Delta S}{(1 + k_d \theta_s)}$

VSS derived from microbial demise (MD): $\text{AMB}(1 - f_b)k_d\theta_s$

where ΔS is the substrate removed (as COD) and f_b is the biodegradable fraction of the active biomass.

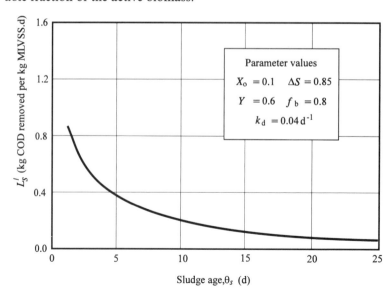

Figure 12.2 Correlation of sludge loading rate and sludge age

The VSS yield is therefore

$$\text{VSS}_y = X_0 + AMB + MD \quad \text{kg VSS per kg COD removed} \quad (12.1)$$

While the rate of COD removal per unit VSS mass is:

$$L'_s = \frac{1}{(X_0 + AMB + MD)\theta_s}$$

which, written in process coefficient terms, becomes:

$$L'_s = \frac{1}{\theta_s(X_0 + Y\Delta S(1 + (1 - f_b)(k_d\theta_s))/(1 + k_d\theta_s))} \quad \text{kg COD} \quad (12.2)$$
$$\text{removed per kg MLVSS} \cdot \text{d}$$

L'_s is plotted as a function of θ_s in Figure 12.2, using the typical characteristic values for municipal wastewater presented in Table 12.1.

Table 12.1 Typical municipal wastewater characteristics (these values are used in subsequent process computations) Concentrations (mg l^{-1})

Condition	Inert SS	VSS	TSS	BOD$_5$	COD	TKN	TP
Raw	50	130*	180	200	400	36	8
Settled	12	60**	72	120	240	32	7

* non-biodegradable fraction 30%
**non-biodegradable fraction 20%
COD of VSS = 1.5 VSS, hence $\Delta S = 1 - 1.5X_0$

For typical municipal sewage treatment applications, the correlation of the applied COD loading rate L_s and L'_s can be approximated as follows:

$$L'_s = L_s(0.76 + 0.004 \ \theta_s) \quad (12.3)$$

12.2 DESIGN CONSIDERATIONS

The basic design parameters for the aeration unit of an AS process are as follows:

(a) the selected value for the sludge loading rate, L_S, or sludge age, θ_s;

(b) the operating value of MLVSS;

(c) the dissolved oxygen (DO) concentration in the mixed liquor.

L_S is more frequently used than θ_s, as the key design parameter, for the very practical reason that its use allows the determination of the required AS process biomass without the need to make reference to the process kinetic coefficients, for which reliable values are not generally available. It has also been the general design practice to use BOD_5 as a design parameter in preference to COD.

The design volume V of the aeration unit follows from the design values chosen for L_S and MLVSS:

$$V = \frac{BOD_5 \text{ load}}{L_s \cdot \text{MLVSS}} \qquad (12.4)$$

where the BOD_5 load is in kg BOD_5 d^{-1}. The product Ls × MLVSS is called the organic space-loading rate, denoted as $L_o(kgBOD_5 \text{ m}^{-3} \text{ d}^{-1})$.

Typical value ranges for the main design parameters are given in Table 12.2

Table 12.2   Value ranges for activated sludge design parameters

Process Category	L_S		θ_s (d)
	kg BOD_5 per kg MLVSS·d	kg COD per kg MLVSS·d	
High rate	>1.0	>1.6	<0.5
Medium rate	0.2–0.4	0.3–0.6	4–8
Low rate	≤0.15	≤0.25	≥10

Although a considerable amount is known about the mechanism and kinetics of the AS process, as discussed in Chapter 11, the assessment of the limiting values which may be assigned to the main design parameters to achieve a particular level of performance in the purification of a particular wastewater is most reliably made through pilot plant operation. Where such pilot plant tests are used as an aid to prototype reactor design it is important to ensure that the prototype environmental conditions are accurately modelled at the pilot plant scale. In a plug-flow reactor, for example, the pilot plant and prototype should have the same substrate gradient between inlet and outlet ends.

While the treated effluent quality is always a key consideration in the design of activated sludge processes, other operational factors also influence the choice of sludge loading rate or sludge age. The more important of these are the degree of stabilisation of the sludge biomass to be achieved and the requirement to produce a sludge biomass with good settling properties (see section 12.5). Only plants of the extended aeration category achieve the degree of stabilization necessary for environmental disposal of the sludge biomass without giving rise to the risk of odour nuisance.

The reported BOD removal performance of a number of pilot and full-scale plants (Downing et al., 1965) as a function of sludge loading rate, is illustrated in Figure 12.3. These data are based largely on results obtained from plants treating municipal wastewaters, which can show considerable variation depending the proportion of industrial waste they contain. This factor, together sludge settleability variation, are

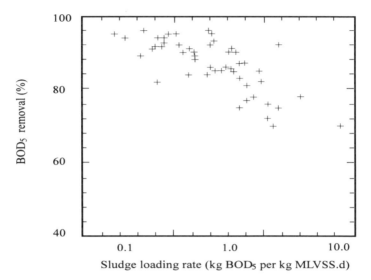

Figure 12.3 Field results showing the influence of sludge loading rate on BOD$_5$ removal from municipal wastewaters (after Downing, 1966)

probably the main reasons for the degree of scatter in the performance levels achieved.

12.3 OXYGEN REQUIREMENTS

A concentration of dissolved oxygen (DO) must be maintained in the mixed liquor to satisfy microbial requirements. For carbonaceous oxidation the limiting concentration is considered to be about 0.5 mg l^{-1}. Nitrifying processes require a somewhat higher DO, as discussed in section 12.7. For process economy it is desirable that the operating level of dissolved oxygen should be as low as possible since the rate of oxygen transfer by aeration systems is proportional to the oxygen saturation deficit and hence energy requirements for aeration are minimized by operating at the lowest satisfactory oxygen concentration.

The required specific oxygen input (OI) can be estimated as the difference between the COD removed and the residual COD of the volatile solids yield of the process, VSS$_y$:

$$OI = 1 - 1.4 \, VSS_y \quad kg \, O_2 \text{ per kg COD removed} \quad (12.5)$$

based on the assumption that the COD of the VSS = 1.4 VSS.

Using the relationships of equations (12.1), (12.2) and (12.3), OI is plotted as a function of sludge loading rate in Figure 12.4. It will be noted that OI decreases with increasing sludge loading rate (reducing sludge age). This reflects a reduced oxygen uptake due to endogenous respiration as the sludge age is reduced and the degree of mineralization and stabilization of the sludge is at a correspondingly lower level.

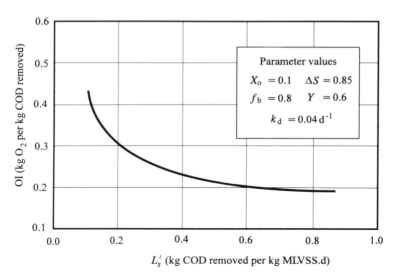

Figure 12.4
Influence of sludge loading on required oxygen input (carbonaceous demand only)

It is important to note that the carbonaceous respiration demand will only represent part of the total demand in activated sludge processes where nitrification is also taking place (see section 12.8), as is likely to be the case at operating L_s values less than about 0.2. The oxygenation capacity of the aeration system must be sufficient to meet the *total* respiration requirements of the process, while at the same time maintaining a specified residual dissolved oxygen (DO) concentration. In normal process operation, the DO level is typically maintained in the range 1–2 mg l^{-1} for carbonaceous oxidation (see section 12.8 for nitrification DO requirements). In determining the required capacity of the aeration system, account must be taken of the diurnal variation in the input BOD and nitrogen loads, which in turn lead to a corresponding

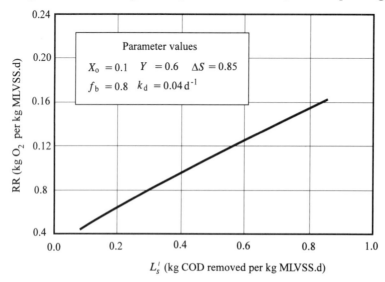

Figure 12.5
Influence of sludge loading on mixed liquor respiration rate

variation in the respiration rate. In plug-flow activated sludge processes, the design of the aeration system must take into account the spatial as well as the temporal variation of the respiration rate in the reactor.

It is also useful to present the oxygen requirement in terms of the sludge respiration rate (RR), expressed as kg O_2 per kg MLVSS.d, which can be correlated with sludge age or with sludge loading rate. Figure 12.5 plots RR as a function of sludge loading rate, L'_s.

12.4 SLUDGE GROWTH

As shown in Section 12.2, the VSS production can be considered to made up of three components:

(1) The influent non-biodegradable VSS, denoted as X_O.

(2) The active microbial biomass, denoted as AMD.

(3) The organic non-biodegradable residue from microbial decay, denoted as MD.

The net yield of VSS (kg VSS per kg COD removed, denoted as VSSy) is the sum of these:

$$\text{VSS}_y = X_o + [Y\,\Delta S(1 + (1 - f_b)(k_d\theta_s))]/(1 + k_d\theta_s) \qquad (12.6)$$

where the process parameters are as defined in section 12.2.

Equation (12.6) indicates that the net sludge production rate is reduced as the sludge residence time θ_S (sludge age) is increased, i.e. high-rate processes produce more microbial biomass residue than extended aeration processes. In addition to the VSS production, in estimating the actual production of excess sludge account must also be taken of the contribution of the inert suspended material in the influent and the loss of suspended material discharged with the effluent. The estimated VSS sludge production, in accordance with equation (12.6), is plotted in Figure 12.6 as a function of sludge age.

Most reported values of excess sludge production rates are within the range 0.35–1.15 kg per kg of BOD_5 removed (0.2–0.6 kg per kg COD removed), the first-mentioned value referring to the upper end of the extended aeration scale. It is worthy of note that in the latter category of plant the sludge lost in the effluent in normal operation may be a significant proportion of total sludge production.

12.5 PHYSICAL CHARACTERISTICS OF AS

12.5.1 Settleability

The most important physical characteristic of an activated sludge is its separability from the liquid in which it is dispersed. Activated sludges

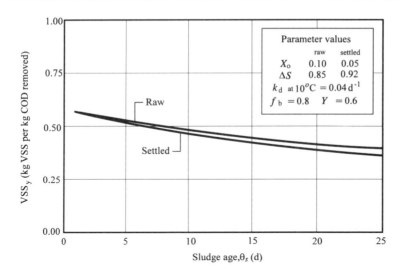

Figure 12.6
Estimated activated sludge VSS yield as a function of sludge age, including non-biodegradable influent VSS

with good separation characteristics are flocculent and settle as shown on Figure 2.3, leaving a clear supernatant free of visible suspended matter. The settleability of an activated sludge is measured by its sludge volume index (*SVI*), which is the volume in ml occupied by 1 g of sludge when 1 l of mixed liquor has been allowed to settle for 30 min in a 1 l measuring cylinder. Normal flocculent activated sludges have *SVI* values within the range 80–120 ml g^{-1}, while non-flocculent or *bulking* sludges may have *SVI* values in excess of 200.

Settleability may also be measured in terms of the stirred sludge volume index (*SSVI*), which is the volume as ml g^{-1} occupied by an AS sample after 30 min settling in a cylindrical vessel, 0.5 m high, stirred at a speed of 1 rpm. It has been found that the *SSVI* test procedure gives better reproducibility than the *SVI* test procedure.

It is generally considered that the floc-forming properties of an activated sludge are related to the morphology of its dominant organism type. Where cocoid and rod-shaped bacteria predominate, activated sludges invariably exhibit good floc-forming properties and hence good settling behaviour. It is considered that the presence of some filamentous bacteria within the floc is beneficial in strengthening the floc structure. However, the predominance of filamentous bacteria is invariably detrimental to floc structure and settling behaviour. The most common detrimental effect of excessive filamentous growth is the creation of so-called bulking sludge, which, as the name implies, has a low aggregative capacity and exhibits poor settling behaviour. About 25 different filamentous organisms are known (Gray, 1990) to cause sludge bulking. The environmental conditions that give rise to the development of a bulking sludge and the commonly associated filamentous organisms are summarized in Table 12.3.

While plant operating experience and research (Pipes, 1967; Chambers and Tomlinson, 1982; Strom and Jenkins, 1984) have shown that the environmental conditions listed in Table 12.3 are favourable to the growth of filamentous organisms, they are not necessarily the only causative factors and may not always give rise to a bulking sludge, the precise cause of which can sometimes be difficult to identify. Bulking occasionally occurs in the absence of filamentous bacteria. This type of bulking behaviour is known as zoogleal or viscous bulking, resulting in a mixed liquor having a slimy non-flocculent consistency (Pipes, 1979).

Table 12.3 Bulking activated sludge

Suggested causative condition	Indicative filament type
Low dissolved oxygen	type 1701, *Sphaerotilus natans, Haliscomeno-bacter hydrossis, Microthrix parvicella.*
Low L_s	*M. parvicella, H. hydrossis, Nocardia amarae,* types 021N, 0041, 0675, 0092, 0581, 0961, 0803.
Septic wastewater/sulphide	*thiothrix, Beggiatoa,* type 021N
Nutrient deficiency	*thiothrix, S. natans,* type 021N
Low pH	Fungi

Source: Strom and Jenkins (1984).

A second detrimental operational characteristic which may result from excessive filamentous growth is the problem of *foaming* in the aeration tank, which is particularly associated with the *Nocardia* and *Microthrix* species (Foot, 1992). Biological foaming takes the form of a floating layer of activated sludge of mousse-like consistency, containing entrapped air bubbles in a mainly filamentous floc structure. It would appear that foam-forming filamentous organisms possess hydrophobic properties and may also be capable of generating extracellular surfactants, both of which are prime ingredients in the production of a floating activated sludge foam layer. Foaming problems appear to be mainly associated with low loading conditions (i.e. low L_s). A variety of operational measures have been applied in the control of foaming (Tipping, 1995; Pitt and Jenkins, 1990), the most common being the reduction of the microbial solids residence time or sludge age, which can be achieved by reducing the operating *MLSS* concentration. Other methods include the use of water sprays, anti-foaming agents, biocide addition and physical removal of the float layer.

As noted in Table 12.3, nutrient deficiency is a potential cause of excessive filamentous growth. The incoming wastewater must provide all the essential nutrients for microbial growth, including nitrogen, phosphorus and trace elements (see Table 11.1). Municipal wastewaters generally satisfy this requirement but some industrial wastewaters may not. Pipes (1979) recommends a BOD_5:N:P ratio of 100:5:1 to prevent bulking. The BOD_5:N:P ratio in domestic sewage is typically in the region of 100:20:3; hence, in most activated sludge applications, there is an excess of nitrogen and phosphorus over that required for microbial metabolism.

12.5.2 Selector tanks

Bulking caused by a low L_s value is most commonly associated with completely mixed extended aeration activated sludge processes. In such reactors there is a uniformly low substrate concentration throughout the reactor volume, which favours the growth of filamentous organisms over floc-formers. Under conditions of high soluble substrate concentration, however, it is believed that the floc-forming organisms can absorb substrate more rapidly than the filamentous species and hence can outgrow the latter. This negative operational characteristic of low-rate completely mixed systems can be neutralized by the use of an upstream selector tank (Chudoba *et al.*, 1973), which is operated at a sufficiently high L_s value to favour the growth of floc-forming organisms over filamentous organisms. Operating L_s values for selector tanks may vary in the range 0.3–0.5 kg BOD_5 per kg MLSS·d. A rapid absorption of soluble substrate by floc-formers takes place in the selector, enabling them to sustain a continuing growth in the following main aeration basin.

It is worthy of note that the foregoing environmental requirement in respect of substrate concentration is automatically satisfied in reactors of plug-flow configuration, in which the substrate concentration decreases from a maximum at the inlet end to a minimum at the outlet end.

12.6 SEPARATION AND RECYCLING OF MIXED LIQUOR

The maintenance of a steady-state mixed liquor biomass concentration is conventionally achieved by recycling the settled sludge from a downstream sedimentation process (secondary sedimentation), as illustrated in Figure 12.1. The secondary sedimentation tank has to perform two functions: (a) it must produce a well-clarified effluent and (b) it must have a sufficient solids flux capacity to allow the sludge to be recycled. The latter is frequently the governing performance criterion.

White (1975) suggested a design procedure for secondary sedimentation tanks, which is based on an empirical correlation of the limiting solids flux capacity, F_L, the sludge underflow rate, Q_u and the sludge $SSVI$:

$$F_L = 307(SSVI)^{-0.77}(Q_u/A)^{0.68} \text{ kg m}^{-2} \text{ h}^{-1} \qquad (12.7)$$

where A is the plan area of the sedimentation tank (m^2) and Q_u is the sludge underflow rate (m^3 h^{-1}).

The specific applied solids loading rate, F_a, is given by

$$F_a = (Q_u + Q)MLSS/A \tag{12.8}$$

where Q is the wastewater flow ($m^3\,h^{-1}$).

Neglecting the leakage of solids in the clarified effluent, it is clear that the solids flux capacity of the tank is reached when $F_a = F_L$. Thus, by combining equations (12.7) and (12.8), the following parametric relationship applies at the limiting solids loading condition:

$$MLSS = 307(SSVI)^{-0.77}\left(\frac{Q}{A}\right)^{-0.32}\left(\frac{R^{0.68}}{1+R}\right) \tag{12.9}$$

where R is the sludge recycle ratio (Q_u/Q) and Q/A is the tank hydraulic surface loading rate ($m\,h^{-1}$).

Equation (12.9) correlates four parameters, $MLSS$, $SSVI$, Q/A and R, at the limiting solids flux condition. This correlation is illustrated graphically in Figure 12.7.

Typically, secondary sedimentation tanks are designed on the basis of a surface loading rate (Q/A) of 1 m h^{-1} and a sludge recycle ratio R of unity. The operational limits relating to these design values can be examined by reference to Figure 12.7. For example, if the operating $MLSS$ is 4 kg m^{-3}, the solids flux capacity of the unit is reached at an $SSVI$ value of 116 ml g^{-1}. If the operating $MLSS$ is reduced to 3 kg m^{-3}, the $SSVI$ may rise to 168 ml g^{-1} before the solids flux capacity is breached. It is clear that if the solids flux capacity is being exceeded due to an increasing $SSVI$ value, the situation may be remedied by increasing the sludge recycle rate or by reducing the $MLSS$ concentration. At high operational $MLSS$ values (3.5 kg m^{-3}), it is clear from

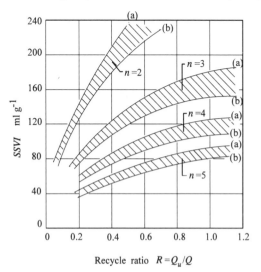

Figure 12.7
Solids flux
capacity in
secondary
sedimentation:
correlation of
process variables
at solids flux limit

Figure 12.7 that there is little solids flux capacity gain by increasing the sludge recycle ratio beyond 1.2.

It should be noted that not all activated sludges behave exactly as predicted by equation (12.9), which has been shown by field observations to have an accuracy of ±20%.

The foregoing discussion of secondary sedimentation makes no reference to tank depth. The depth must be sufficient to accommodate a clear water zone and a sludge-blanket zone. In practice, design depths in the range 2–3 m have been found to be adequate.

Dewaterability is also an important characteristic of sludges, particularly in relation to handling and disposal. In general, activated sludges with good settling characteristics (good floc-forming properties) will also have good dewatering characteristics (see to chapter 16 for further discussion of sludge dewatering).

12.7   NITROGEN REMOVAL

The biochemical removal of nitrogen from wastewaters is carried out in two distinct process stages. The first stage is the process of *nitrification*, or the conversion of ammonia to nitrate, and the second stage is *denitrification*, or the reduction of nitrate to gaseous nitrogen endproducts.

12.7.1   Nitrification

Microbial nitrification is a two-step process, the first step being the conversion of ammonia to nitrite, which is accomplished by *Nitrosomonas* bacteria, while the second step involves the conversion of nitrite to nitrate by *Nitrobacter* bacteria:

The overall chemical oxidation reaction is

$$NH_4^+ + 2O_2 \rightarrow NO_3^- + 2H^+ + H_2O$$

Taking account of the incorporation of nutrients in the process of cell synthesis, using yields of 0.08 g VSS per g NH_4^+–N and 0.05 g VSS per g NO_2^-–N, for *Nitrosomonas* and *Nitrobacter*, respectively (USEPA, 1991), the overall reaction describing the complete nitrification process becomes:

$$1.00NH_4^+ + 1.89O_2 + 0.0805CO_2 \rightarrow 0.0161C_5H_7O_2N$$
$$+ 0.952H_2O + 0.984NO_3 + 1.98H \tag{12.11}$$

where $C_5H_7NO_2$ is taken as representing the bacterial cell composition. Thus the conversion of 1mg of ammonium-nitrogen is estimated to result in the consumption of 4.32 mg oxygen, the production of 0.13 mg of nitrifying organisms and the destruction of 7.07 mg of alkalinity (as $CaCO_3$).

Nitrification Kinetics

The nitrifying bacteria are chemoautotrophs; their growth energy is derived from the oxidation of inorganic nitrogen and their carbon source is carbon dioxide.

The growth rate of nitrifiers is estimated to be some 10–20 times slower than the growth rate of the heterotrophs which are responsible for carbonaceous BOD removal. Of the two species responsible for nitrification, *Nitrobacter* has a higher growth rate than *Nitrosomonas*. The growth rate of the latter, which is responsible for the conversion of ammonia to nitrite, is thus normally rate-limiting for the nitrification process. It also follows from this that nitrite is not usually found in high concentrations in nitrifying processes operating under steady state conditions. The growth of *Nitrosomonas* can be expressed according to the Monod growth model (see chapter 11, equation (11.2)) as follows:

$$\mu_N = \hat{\mu}_N \frac{N}{K_N + N} \qquad (12.12)$$

where

μ_N = specific growth rate of *Nitrosomonas* (d^{-1})

$\hat{\mu}_N$ = maximum specific growth of *Nitrosomonas* (d^{-1})

K_N = half-saturation coefficient for *Nitrosomonas* (mg l^{-1} NH$_4^+$–N)

N = NH$_4^+$–N concentration (mg l^{-1})

For design purposes, the value of K_N may be taken as 1 mg NH$_4^+$–N l^{-1}, while the value of the maximum specific growth rate constant is dependent on temperature and may be represented by the following empirical Arrhenius-type expression (USEPA, 1993):

$$\hat{\mu}_N = 0.47 \, \theta^{(T-15)} \qquad (12.13)$$

where θ is generally taken to have a value of about 1.1.

Because of the relatively low value of K_N, the nitrification process proceeds, under typical operating wastewater treatment conditions, at the maximum growth rate for the *Nitrosomonas* bacteria, i.e. it is a zero-order process, independent of the ammonia concentration. If, however, the ammonia nitrogen concentration drops towards the half saturation concentration level of 1 mg l^{-1}, then the process becomes rate-limited by the reduced concentration according to equation (12.12).

As previously shown (Chapter 11, equation (11.14)), the effluent substrate (in this instance NH$_4^+$–N) in a completely mixed reactor, operating at steady state, can be expressed in terms of the sludge age and the Monod kinetic parameters, as follows:

$$NH_4^+ - N = \frac{K_N(1 + k_d\theta_s)}{\theta_s(\hat{\mu}_N - k_d) - 1} \qquad (12.14)$$

This relationship is plotted in Figure 12.8 for temperatures of 10°C and 20°C, illustrating the marked influence of temperature on nitrification performance. For example, Figure 12.8 indicates that a sludge age of about 10 d is necessary to reduce the reactor ammonia nitrogen concentration to 1 mg l^{-1} at an operating temperature of 10°C, while at a temperature of 20°C the ammonia nitrogen concentration is reduced to 0.5 mg l^{-1} at an operating sludge age of about 5 days.

It has been observed (Stenstrom and Song, 1991) that the *dissolved oxygen* concentration (DO) has a significant influence on the nitrification process. It would appear that the growth rate of *Nitrosomonas* may be slowed down at DO concentrations less than 1 mg l^{-1}. The achievement of this limit value throughout the mixed liquor biomass, however, may require an operating DO level of at least 2 mg l^{-1}, depending on the intensity of mixing and the associated spatial DO gradients in the microbial floc.

As noted above, the nitrification process exerts a substantial alkalinity demand (7.1 mg alkalinity as CaCO3 per mg NH$^+$–N). This inevitably reduces pH, particularly where there is an insufficient buffering capacity in the wastewater being treated. The optimum pH range for nitrification would appear to be 7.0–8.5. For design purposes, this range may be extended to 6.5–9.0 (USEPA, 1993).

Nitrifying organisms are susceptible to inhibition by many organic substances, such as solvents, and by inorganic substances, such as heavy metals (Hockenbury and Grady, 1977; Benmoussa *et al.*, 1986). It has also been reported (Gujer, 1977) that the recycling of anaerobic digester supernatant may have an inhibitory effect on *Nitrosomonas* growth rate.

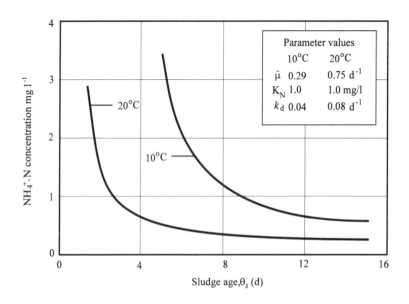

Figure 12.8
Steady-state ammonia nitrogen concentration in a completely mixed nitrification reactor

12.7.2   Denitrification

The microbial reduction of nitrate is brought about by a variety of oxygen-utilizing heterotrophic bacteria which, in the absence of oxygen, are capable of using nitrate in place of oxygen as a terminal electron acceptor. Research has shown that anoxic respiration of this kind also takes place at low DO concentration; the DO concentration at which denitrification stops has been reported to be 0.2 mg l^{-1} in pure cultures (Focht and Chang, 1975) and in activated sludge systems to be in the range 0.3–1.5 mg l^{-1} (Burdick *et al.*, 1982). In the denitrification process, nitrate and its reduced forms act as electron acceptors, resulting in a stage-wise reduction of nitrate to gaseous nitrogen:

Redox state of nitrogen:	+5	+3	+2	+1	0
	NO_3^-	NO_2^-	NO	N_2O	N_2
	nitrate	nitrite	nitric oxide*	nitrous oxide*	nitrogen*

* gaseous end products

A variety of heterotrophic bacteria, commonly present in activated sludges, participate in the denitrification process including *Alcaligenes*, *Achromobacter*, *Micrococcus* and *Pseudomonas*. These organisms have the remarkable metabolic capability of using either oxygen or nitrate as an electron acceptor in their energy generation process. If oxygen is present, it is preferentially used over nitrate.

As heterotrophs, these bacteria use organic matter as their carbon source and hence remove BOD in conjunction with denitrification. The overall stoichiometry, neglecting cell synthesis, may be approximated as follows:

$$2NO_3^- + 2H^+ + 2.5C \rightarrow N_2 + 2.5CO_2 + H_2O \qquad (12.15)$$

Based on electron acceptor capacity, 1 g of nitrate nitrogen is equivalent to 2.86 g of oxygen. Thus, the combination of nitrification and denitrification processes can significantly reduce the overall oxygen demand relative to nitrification on its own.

The organic carbon requirement for the denitrification process, expressed in BOD terms, corresponds to a BOD_5 per NO_3^-–N ratio $\geqslant 3$. Thus, a typical fully nitrified effluent will not have sufficient residual biodegradable carbon to act as a carbon source for the denitrification process. In practice, either the wastewater influent or a supplemental source such as methanol is used to provide the necessary organic carbon for the denitrification process.

The denitrification process increases the bicarbonate alkalinity, theoretically creating 3.57 mg alkalinity as $CaCO_3$ per mg of nitrate nitrogen reduced to nitrogen gas. This recovery of alkalinity partially reverses the drop in alkalinity and pH associated with the preceding nitrification process. This compensatory effect can be a significant process design consideration for wastewaters that are low in alkalinity and may be sufficient to prevent an inhibitory drop in pH in the nitrification step.

Denitrification Process Kinetics

The rate of growth of denitrifying organisms can be expressed in a Monod-type expression, using nitrate as the rate-limiting nutrient:

$$\mu_D = \hat{\mu}_D \frac{D}{K_D + D} \qquad (12.16)$$

where

μ_D = specific growth rate (d^{-1}),
$\hat{\mu}_D$ = maximum specific denitrfier growth rate,
D = concentration of nitrate nitrogen (mg l^{-1}),
K_D = half saturation coefficient (mg l^{-1})

Since the value of K_D is very low (0.3–0.6 mg l^{-1} NO$_3$–N, see Table 11.3), the process is effectively zero order with respect to nitrate concentration.

The other material environmental influences are the concentrations of biodegradable carbon and of DO, the latter having an inhibiting effect, as discussed above. These influences can be combined with (12.16) to give the following overall expression for denitrifier growth rate:

$$\mu_D = \hat{\mu}_D \left(\frac{D}{K_D + D} \right) \left(\frac{S}{K_s + S} \right) \left(\frac{K_o}{K_o + S_o} \right) \qquad (12.17)$$

where S and K_s refer to the biodegradable organic substrate and S_o and K_o refer to DO. The K_s value may be taken as that which applies to heterotrophic growth under aerobic conditions.

The term $K_0/(K_0 + S_0)$ acts as a switching function, turning the denitrification process on and off. A value of K_0 of 0.1 mg l^{-1} has been suggested (IAWPRC model, 1994) for denitrification, implying a halving of the growth rate at a DO concentration of 0.1 mg l^{-1} relative to its value at zero DO.

As with all biological processes, denitrification is significantly influenced by temperature. The magnitude of this influence can be expressed by an Arrhenius-type function of the form:

$$\mu = \mu_{20} \, \theta^{(T-20)} \qquad (12.18)$$

where values of θ have been reported (USEPA, 1993) in the range 1.02–1.08. At a θ value of 1.05, the mean of this range, the growth rate at 10°C, is calculated to be 61% of its rate at 20°C.

In general, the denitrification process is much less sensitive to inhibitory substances than is the nitrification process. Experimental findings

(USEPA, 1993) indicate that denitrification rates may be depressed below pH 6.0 and above pH 8.0.

12.7.3 Design of nitrogen-removal systems

It is clear from the foregoing discussion that biological nitrification and denitrification processes have conflicting environmental requirements for optimal operation. Nitrification requires a highly aerobic environment with a sufficiently long microbial residence time to allow the development of a sufficiently high concentration of the slow-growing nitrifying bacteria, *Nitrosomonas* and *Nitrobacter*. These conditions result in a very low biodegradable carbon substrate level in nitrifying reactors. Denitrification, on the other hand, requires an anoxic environment and the availablility of an ample biodegradable carbon substrate concentration ($BOD_5 : NO_3^--N \geqslant 3$).

Nitrification

The minimum sludge age for nitrification corresponds to the inverse maximum growth rate:

$$\theta_{s(min)} = \frac{1}{\hat{\mu}_N}$$

where $\hat{\mu}_N$ is a function of temperature, as defined by equation (12.13). One approach to selecting the design sludge age, θ_{sd}, is to apply a design service factor (DSF) to $\theta_{s(min)}$ to take into account fluctuations in load and the desirability of operating in a stable nitrification zone (see Figure 12.8):

$$\theta_{sd} = \frac{DSF}{\hat{\mu}} \tag{12.19}$$

For example, at an operating temperature of 10°C and using a DSF value of 3.0, the value of θ_s is found from Equations 12.13 and 12.19 to be 10.2 days. The corresponding COD removal rate at this sludge age is found from Figure 12.2 to be about 0.25 kg per kg MLVSS per day. This latter value, together with the magnitude of the total COD load, is then used to compute the required total mass of MLVSS and hence the reactor volume is determined based on a selected operating MLVSS concentration.

In situations where the COD:N ratio is high, it may be economically worthwhile to consider a two-stage biological process, the first stage being operated at a non-nitrifying sludge age and hence removing most of the carbonaceous BOD at an efficiently high specific conversion rate in a correspondingly low process volume. The second stage can then be designed as a nitrification stage, resulting in a significant reduction in the overall process volume relative to the volume required for a single-stage process.

Denitrification

A denitrification process requires an adequate carbon source to sustain the growth of the denitrifiers and a sufficient residence time under anoxic conditions to allow microbial utilization of the nitrate. The rate of growth of the denitrifiers (equation. (12.17)) is inhibited by dissolved oxygen and positively influenced by the organic substrate and nitrate concentrations. Where denitrification processes are accompanied by nitrification processes, the rate of respiration in the latter may be used as the basis for estimating the nitrate uptake rate in the denitrification process. The rate of respiration, RR, is plotted as a function of the sludge loading rate L_s in Figure 12.5. The required MLVSS biomass (M_D) for dentrification is then calculated from the following relation:

$$M_D = \frac{2.86 \Delta N}{RR} kg \qquad (12.19)$$

where ΔN is the nitrate reduction rate (kg NO_3^-–N d^{-1}), the multiplier 2.86 converts nitrate to the equivalent oxygen, and RR is the aerobic respiration rate (kg O_2 (kg MLVSS d)$^{-1}$). The use of RR, the nitrification respiration rate, provides a margin of safety in the denitrification reactor sizing, since the organic substrate is always higher in the denitrification reactor than in the nitrification reactor and hence the heterotrophic growth rate should also be correspondingly higher.

Combined Nitrification and Denitrification Processes

Biological nitrogen removal requires that nitrification and denitrification processes are combined in an integrated treatment system. Such an integrated system is illustrated in Figure 12.9. In this layout the influent wastewater is fed to the anoxic reactor, which in turn feeds the aerobic reactor. The nitrified effluent from the aerobic reactor is recycled together with the settled sludge from the downstream sedimentation tank to the inlet of the anoxic reactor. This anoxic/aerobic sequence uses the influent organic substrate as the carbon source for the denitrification process and obviates the need for a supplementary carbon

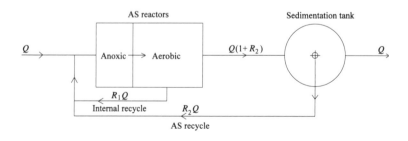

Figure 12.9
Combined nitrification/ denitrification AS process schematic

feed, which is the case when the reactor order is reversed. However, the anoxic/aerobic order requires a significant total recyle rate if a low effluent nitrate concentration is to be achieved. Assuming that both reactors are completely mixed and operating at a steady state, such that the nitrate nitrogen concentration in the anoxic reactor is C_N and the ammonia nitrogen concentration in the aerobic reactor is C_{AE} (also the effluent ammonia nitrogen concentration), then the effluent nitrate nitrogen concentration C_{NE} is a function of the recycle rate as follows:

$$C_{NE} = C_N + \frac{C_{AI}(1 - K) - C_{AE}}{1 + R} \tag{12.20}$$

where C_{AI} is the influent TKN concentration and the factor K relates to the incorporation of nitrogen into the cell biomass; R is the overall recycle ratio, i.e. $(R_1 + R_2)$. The factor K is quantified on the basis the of a mean N:COD ratio of 0.06 in the MLVSS and a VSS yield:COD removed ratio of 0.4:

$$K = 0.024 \frac{COD_r}{TKN} \tag{12.21}$$

where COD_r is the removed COD.

Combined Nitrification/Denitrification Design Example

Data:

mean daily wastewater flowrate $= 3000 \text{ m}^3 \text{ d}^{-1}$;

influent COD $= 400 \text{ mg l}^{-1}$;

influent TKN $= 36 \text{ mg l}^{-1}$;

process design temperature $= 10°C$;

influent VSS $= 130 \text{ mg l}^{-1}$ (30% non-biodegradable)

(note that the foregoing process design graphs have been computed on the basis of this wastewater composition);

Target effluent $NH_3^+-N = <1 \text{ mg l}^{-1}$;

target effluent $NO_3^--N = 5 \text{ mg l}^{-1}$.

Assume 80% COD removal:

COD removed $= 3000 \times 0.400 \times 0.8 = 960 \text{ kg d}^{-1}$.

Aerobic reactor: $\theta_{sd} = \dfrac{DSF}{\hat{\mu}} = \dfrac{3}{0.47(1.1)^{-5}} = 10.3\text{d}$

From Figure 12.2, the corresponding value of $L_s' = 0.2 \text{ kg COD}_r \text{ (kg MLVSS d)}^{-1}$

Assuming an MLVSS of 2.5 kg m^{-3},

$$\text{aerobic reactor volume} = \frac{960}{0.2 \times 2.5} = 3840 \text{ m}^3$$

At a sludge age of 10.3d the effluent NH_3^+-N is found from:

$$\mu_N \frac{N}{K_n + N} - k_d = \frac{1}{\theta_s} \Rightarrow 0.29\left(\frac{N}{1 + N}\right) - 0.04 = \frac{1}{10.3}$$

Hence
$N = 0.24$ mg l^{-1} (< 1 mg l^{-1}, as required)

Denitrification: nitrogen incorporation factor (equation (12.21)
$K = 0.024 \times 320/36 = 0.21$

$$\Delta N = 36(1 - 0.21) - (0.24 + 5) = 23.2 \text{ mg l}^{-1} \Rightarrow 69.6 \text{ kg d}^{-1}$$

From Figures 12.2 and 12.5: respiration rate at sludge age of 10.3
d $= 0.062$ kg O$_2$ (kg MLVSS d)$^{-1}$

From equation (12.19): required anoxic reactor volume $= \dfrac{69.6 \times 2.86}{0.062 \times 2.5}$
$= 1284$ m^3

Ratio of anoxic:aerobic volume $= \dfrac{1284}{3840} = 0.33$

For typical domestic wastewater (BOD$_5$/TKN in the range 4–5), the anoxic reactor volume for denitrification may be about 30% to 50% of that required for nitrification. The anoxic zone requires an adequate intensity of mixing to prevent sedimentation of the activated sludge.

12.8 BIOLOGICAL PHOSPHORUS REMOVAL

The activated sludge process can be manipulated to enhance phosphorus-removal through creating process environmental conditions that produce the so-called 'luxury' uptake of phosphorus, i.e. an uptake in excess of the normal metabolic fixation of phosphorus by bacterial cells. This enhanced uptake results in the phosphorus-content of the biomass being increased from a typical 1.5–2.0% on a dry weight basis in a conventional AS process to the region of 3–6% in a biological phosphorus removal process (Stensel, 1991).

This enhanced uptake of phosphorus is achieved by subjecting the mixed liquor to an anaerobic/aerobic cycle. In the anaerobic phase there is an uptake of fermentation products (volatile fatty acids (VFAs) such as acetic acid and propionic acid), which accumulate as storage products within the microbial cells. This uptake of VFAs is accompanied by a corresponding release of cell phosphorus into solution. In the following aerobic stage the stored products are oxidized, resulting in a simultaneous enhanced uptake of phosphorus which is stored as polyphosphate within the cell.

The enhanced removal of phosphorus in the activated sludge process is considered to be due to a specific genus of bacteria, *Acetinobacter calcoaceticus*. The amount of phosphorus incorporated into the microbial biomass would appear to be not greatly influenced by temperature in the range 5–20 °C, with some research evidence to indicate a higher incorporation at lower temperatures (Stensel, 1991).

The two main factors that are known to influence biological phosphorus-removal are the VFA:P ratio in the anaerobic reactor and the sludge age in the aerobic reactor. The VFA availability is dependent on the prior fermentation conditions and carbon substrate availability (BOD). Fukase *et al.* (1982) found that the BOD:P removal ratio increased from 19 to 26 as the sludge age was increased from 4.3 to 8.0 days. At the same time the phosphorus-content of the activated sludge decreased from 5.4%–3.7%. In practice, it may be difficult to achieve an effluent total phosphorus (TP) level in the range 1–2 mg l^{-1}, where the BOD_5:TP ratio in the influent is less than 20.

The *anaerobic reactor* is usually sized to provide an hydraulic residence time of 1–2 h, based on process influent flow (USEPA, 1987). The shorter residence time is adequate for septic wastewaters having a relatively high BOD:P ratio, while the longer residence time may be necessary to allow some breakdown of particulate BOD in wastewaters with a low soluble BOD content. As far as possible nitrate should be excluded from the anaerobic reactor, as its presence inhibits the fermentation process that produces VFAs.

The *aerobic* reactor is designed as a conventional activated sludge process, the selected design sludge age depending on whether nitrification is required or not. As a general rule, where the aerobic reactor is designed for nitrification, a denitrifying anoxic zone should also be provided to avoid carryover of nitrate into the anaerobic reactor for the reasons already outlined. As noted above, the incorporated phosphorus content of AS decreases with increasing sludge age as does the process sludge yield. Hence, the required BOD:P ratio for a given residual phosphorus concentration increases with the aerobic reactor sludge age.

In addition to the foregoing process design considerations, two operational considerations, in particular, must be taken into account to achieve a low residual effluent phosphorus concentration: (a) the effluent suspended solids must be maintained at a low level, since they have a high phosphorus concentration and (b) the excess sludge must be maintained in an aerobic state to avoid phosphorus loss.

A process layout for phosphorus removal, incorporating both enhanced biological removal and chemical precipitation by lime, is shown in Figure 12.10. Figure 12.11 demonstrates a process scheme for combined biological removal of nitrogen and phosphorus.

The anaerobic reactor in Figure 12.10 effects a microbial release of cellular phosphorus into the supernatant water which is then transferred to the lime tank to be precipitated as calcium phosphate. Thus, this process arrangement effectively uses microorganisms to transfer phosphorus from the main wastewater stream into a more concentrated side stream and chemical precipitation by lime.

The Bardenpho process combines biological precipitation of phosphorus in the anaerobic reactor at the upstream end of the process

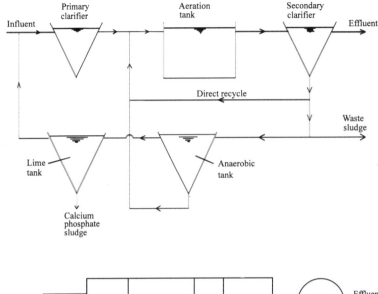

Figure 12.10
Schematic layout of 'Phostrip' process for phosphorus removal. *Source:* (Levin et al., 1975)

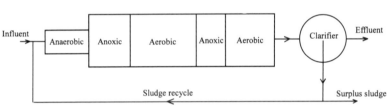

Figure 12.11
Schematic layout of 'Bardenpho' process for nitrogen and phosphorus removal. (Reprinted with permission from Stensel, 1991. © Lewis Publishers, an imprint of CRC Press, Boca Raton, Florida)

stream outlined in Figure 12.11. The aerobic zone is designed to achieve complete nitrification. There is a high internal recirculation between the aerobic and first anoxic zone, where up to 80% of the nitrate is removed. The remainder of the nitrate is removed in the secondary downstream anoxic zone, which is followed by a further aerobic stage prior to sludge separation in a secondary clarifier.

12.9 CHOICE OF PROCESS

Within the spectrum of activated sludge process variations found in design practice, the size chosen for the aeration unit may vary from a volume V in a high rate plant, providing carbonaceous oxidation only, to a volume of about $50V$ in an extended aeration plant such as an oxidation ditch, providing complete nitrification.

High rate plants have the advantage of minimizing space and oxygen input requirements. However, they have the disadvantage of producing unstable effluents, and a large yield of active sludge biomass which may have bulking tendencies. Intermediate rate plants (sludge age 5–8 d), as widely used throughout the world for the treatment of municipal wastewaters, produce a stable effluent, having a BOD_5 value in the range 10–20 mg l^{-1}. The excess sludge produced in these plants is also highly active and requires stabilizing by further treatment before disposal.

Extended aeration (EA) activated sludge processes have the advantage of producing a stable nitrified effluent and a low yield of well-mineralized stable excess sludge, with generally favourable dewatering characteristics (good floc-forming characteristics). They have the disadvantage of requiring a large aeration reactor volume and a high oxygenation capacity (the aeration process has a high energy demand).

The extended aeration process is typically selected for small to medium-sized plants, since it eliminates the need for further sludge stabilization. The oxidation ditch, the general configuration of which is shown in Figure 12.12, is an example of a widely used form of the extended aeration process (Casey and Clerkin, 1969). The reactor is essentially a closed loop tank in which a sufficiently high horizontal flow velocity is generated by the bladed rotor aerator to prevent the microbial biomass from settling. Oxidation ditches are typically designed on the basis of a sludge loading rate of about 0.05 kg BOD_5 per kg MLSS per day (corresponding sludge age 25 d). An upstream primary sedimentation process is usually omitted.

Oxidation ditches are typically designed on the basis of a sludge loading rate of about 0.05 kg BOD_5 per kg MLSS per day (corrsponding sludge age >25 days). An upstream primary sedimentation process is usually omitted.

The so-called *contact stabilisation* process is a variant of the activated sludge process, which has separate aerated contact and stabilisation tanks. The influent wastewater is treated in the contact tank which has a relatively short retention time, while the recycled sludge is stabilised in a separate aeration tank prior to being returned to the contact tank. The contact-stabilisation process offers advantages for the treatment of wastewaters whose organic content is mainly in suspended or colloidal forms. Such organics are readily adsorbed on contact with activated sludge floc and hence a short contact period is adequate (15–120 mins). Since the concentration of sludge in the stabilisation tank is normally at least twice that of the mixed liquor, a given degree of sludge

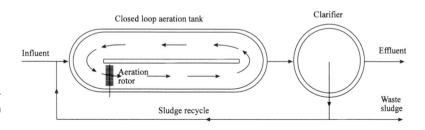

Figure 12.12
Schematic layout of oxidation ditch system (Pasveer (1958))

stabilisation can be effected in a tank of only half the size of that which would be required if the whole of the mixed liquor were aerated. A disadvantage of the system is that only partial removal of slowly oxidisable non-adsorbed dissolved organics is effected in the short contact times used.

REFERENCES

Benmoussa, H, Martin, G. Richard, Y and Leprince, A (1986) Inhibition of nitrification by heavy metal cations, *Water Research.*, **20**, 1333.

Burdick, C. R, Refling, D. R. and Stensel, H.D. (198) Advanced biological treatment to achieve nutrient removal, *JWPCF*, **54**, 1078–1086.

Casey, T. J. and Clerkin, J. P. (1969), Performance characteristics of an oxidation ditch, *Wat. Poll. Cont.*, **68**, No. 6, 687.

Chambers, B and Tomlinson, E. J., eds (1982) *Bulking of Activated Sludge*, Ellis-Horwood, chichester, UK.

Chudoba, J., Grau, P. and Ottova, V. (1973) Control of activated sludge bulking: II Selection of micro-organisms by means of a selector, *Water Research*, **7**, 1389–1406.

Downing, A. L. (1966) Activated sludge plant design problems, *Process Biochemistry*, **2**.

Downing, A. L., Jones, K. and Hopwood, A. P. (1965) Some factors of importance in the design of activated sludge plants, A.I.Ch.E./I.Chem..E. Symposium Series No.9.

Focht, D. D. and Chang, A. C. (1975) Nitrification and denitrification processes in wastewater treatment, *Adv. Appl. Microbiol.* **19**, 153–186.

Foot, R. (1992) The effects of process control parametes on the composition and stability of activated sludge, *J. CIWEM*, **6**, (2), 215–227.

Fukase, T., Shibeta, M. and Mijayi, X. (1982) Studies on the mechanism of biological phosphorus removal, *Japan J. Wat. Poll. Res.*, **5**, 309.

Gray, N. F. (1990) Activated Sludge, Oxford University Press, Oxford.

Gujer, W. (1977) Discussion on the paper dynamic nature of nitrifying biological suspended growth systems, *Prog. Water Tech.*, **9**, 279.

Hockenbury, M. R. and Grady, C. P. L (1977) Inhibition of nitrification – effects of selected organic compounds, *JWPCF*, **49**, 768.

IAWPRC Task Group on Mathematical Modelling for Design and Operation of Biological Wastewater Treatment Processes, Activated Sludge Model No. 1 (1987), Activated Sludghe Model No. 2 (1994), IAWPRC, London.

IWPC (1987) Activatd Sludge, Manual of British Practice in Water Pollution Control, The Chartered Institute of Water and Environmental Management, London, UK.

Levin, G. V., Topol. G. J. and Tornay, A.G.(1975) Operation of a full scale biological phosphorus removal plant, *JWPCF*, **47** (3), 1940.

Pasveer, A. (1958) A simple method of treating small quantities of wasterwater, *Tech. Sanit. Munic.*, **53**, 245.

Pipes, W. O. (1967) Bulking of activated sludge, *Advances in Applied Microbiology*, **9**, 185–234.

Pipes, W. O. (1979) Bulking, deflocculation and pin-point floc, *J. WPCF*, **51**, 62–70.

Pitt, P. and Jenkins, D. (1990) Causes and control of Nocardia in activated sludge, *JWPCF*, **62**, (2), 143–150.

Stensel, H. D. 91991) Principles of biological phosphorus removal, in *Phosphorus and Nitrogen Removal from Municipal Wasterwater*, ed. Sedlack, R. Lewis Publishers, New York.

Stenstrom, M. K. and Song, S. S (1991) Effects of oxygen transport limitation on nitrification in the activated sludge process, *JWPCF*, **63**, 208.

Stewart, M. J. (1964) Activated sludge variations, the complete spectrum, *Water & Sewage Works*, **111**, 4,5, 6–3 parts.

Strom, P. F. and Jenkins, D. Identification and significance of filamentous microorganisms in activated sludge, *J. WPCF*, **56**, 449–59.

Tipping, P. J. (1995) Foaming in activated sludge processes: an operator's overview, *J. CIWEM*, **9**, 281–289.

USEPA (1987) Design Manual: Phosphorus Removal, EPA/62-1/1- 87/001.

USEPA (1993) Nitrogen Control, Technomic Publishing Co., Inc., Lancaster, USA.

White, M. J. D. (1975) Settling of activated sludge, *Tech. Rep.* TR11, WRc.

Related reading

WPCF (1987) Activated Sludge, Manual of Practice OM-9, The Water Pollution Control Federation, Alexandria, VA, USA.

Sedlak, R., ed. 91991) *Phosphorus and Nitrogen Removal from Wastewater*, Lewis publishers, New York.

Horan, N. J. (1990) *Biological Wastewater Treatment Systems*, John Wiley & Sons, Chichester, UK.

13

Aerobic Biofilters

13.1 INTRODUCTION

Aerobic biofilters (also known as 'trickling filters' or 'percolating filters') have been in use for the treatment of organic wastewaters since the first decade of this century. They were originally developed from the practice of sewage treatment by application to land, or 'sewage farming', as it was known in the nineteenth century. Attempts to improve the efficiency of land treatment led to the discovery in about 1888 that settled sewage could be purified by passing it through an artificial bed of coarse porous medium at about 10 times the application rate for conventional land treatment. Following this discovery, percolating filters were developed fairly rapidly and by 1908 this method had become the standard method of accelerated sewage purification. Today, aerobic biofilters compete with the activated sludge process for the purification of dilute organic wastewaters, while high-rate biofilters are sometimes used for the partial treatment of strong organic wastewaters.

Aerobic biofilters may be classified as biological reactors of the attached film / static medium type. The medium most widely used is natural stone (50–100 mm size), as illustrated in Figure 13.1, while a variety of purpose-made plastic media is also available.

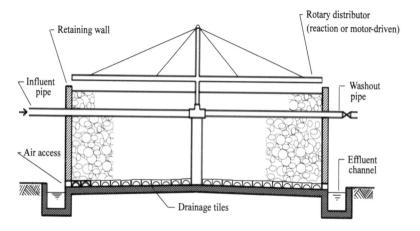

Figure 13.1
Schematic layout of a stone-filled biofilter

13.2   MODE OF ACTION

The mechanism of removal of organics is similar to that of the activated sludge process. The greater portion of the liquid applied to the surface of the filter passes rapidly through and the remainder trickles slowly over the surface of the slime growth. The removal of organic colloids in suspension and dissolved substances occurs by 'biosorption' and coagulation from that portion of the flow that passes through rapidly, and by the usual processes of synthesis and respiration from the part of the flow with long residence time. This residence time is primarily related to the hydraulic loading, so it seems reasonable that the greater the hydraulic loading the more the process will depend upon biosorption and the less it will depend upon synthesis and respiration, other things being equal.

The action of a filter depends on the metabolic activity of zoogleal or filamentous bacteria or of fungi. These colonize the extensive surfaces of the support medium and form the basis of the 'film' which also contains a population of protozoa as well as amorphous solids derived from the waste (see Figure 13.1). Algal growths may also be present in the upper regions of the filter exposed to light. A population of macro-organisms is also usually found associated with the film. The composition of the microbial population depends on the nature of the waste, its strength, the rate at which it is applied and the method of operation of the filter. Therefore the film may range in character from a tenuous bacterial slime in a filter receiving a waste of low organic content, to a thick mass of fungal mycelium in a filter treating a strong sewage containing certain types of industrial wastes.

The reduction in concentration of polluting matter occurs most rapidly in the upper regions of the filter. In a conventional filter treating sewage, about 90% of the biodegradable matter may be removed in the upper 0.6 m of the bed (see Figure 13.2). The net removal rate of organic matter is a function of the immediate removal of the readily biodegrad-

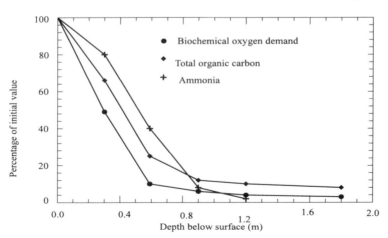

Figure 13.2
Substrate removal over the depth of a biofilter (Reprinted from Bruce 1969 with kind permission from Elsevier Science Ltd, The Boulevard, Langford Lane, Kidlington OX51GB, UK).

able fraction and the release of products of endogenous metabolism of the biofilm. When ammonia is present in the waste, as in the case of sewage, its removal by microbial nitrification occurs largely in the middle and lower regions of the filter. The zones of carbonaceous oxidation and nitrification are not sharply defined, but conditions for the growth of nitrifying bacteria tend to be more favourable in the lower parts of the filter where the content of oxidizable organic matter in the waste has been reduced and where the oxygen concentration in the film is therefore higher (Bruce, 1969).

Mechanical filtration is not an important part of the purification mechanism, so that the term 'filter' is a misnomer persisting from the early days when the nature of the process was misunderstood. Although solids present in the waste may be physically trapped by the medium or the film matrix, it is essential for the satisfactory operation of conventional filters to remove as much coarse suspended matter as possible by primary sedimentation.

As a result of the growth of bacterial cells and of the deposition of coagulated solids from the waste, the quantity of film within the filter tends to increase. The growth rate is dependent on the concentration of degradable organic matter in the waste and on the application rate. The growth of the film tends to be greatest in the upper regions of the filter where the largest proportion of the substrate is removed. If the growth rate of the film exceeds its removal rate by other mechanisms, the interstices of the medium will ultimately become choked, impeding the flow through the bed and restricting ventilation. In the extreme, the surface of the filter may become ponded. This clogging of the medium with accumulated film is the most serious problem in the operation of conventional filter beds and it is the factor that limits the maximum rate at which wastes can be treated efficiently by such plants.

13.3   ECOLOGICAL ASPECTS (Bruce, 1969)

The biological film or slime layer is inhabited by an interdependent microbial population, including bacteria, fungi, protozoa and a variety of macroinvertebrates. Algae are restricted to the surface of the bed, where light is available. They play a minor role in the purification process, but excessive growth can cause blinding of the surface.

The thickness of the film, which can be maintained in an aerobic condition by diffusion of atmospheric oxygen through the film surface, is rather limited. Estimates of the aerobic zone in an actively respiring film vary between 0.06 mm and 2 mm, while in deeper regions of the film anaerobic conditions prevail.

Macroinvertebrates include a variety of worms, insect larvae and snails. They serve a useful role in film disintegration, especially in low-rate filters, where the scouring effect of the liquid flow is rather limited.

One disadvantage of the presence of fly larvae in a filter is the emergence of adult flies, particularly *Psychoda* and *Anisopus*, which may cause a serious nuisance during the warmer months. Various methods (Tomlinson and Jenkins, 1947) for controlling the fly population have been used including treatment of the filter with lime, creosote, bleaching powder and salt. The most effective method of artificial control is to use an efficient insecticide in amounts sufficient to kill the flies and their larvae without affecting the worms and other animals in the filter, and also without rendering the effluent toxic to fish. It has been suggested that the only satisfactory long-term solution to the problem lies in the biological suppression of the fly population by a dominant population of worms competing for the available food supply.

13.4   SEASONAL VARIATIONS

Variations in temperature have a greater influence on the behaviour and performance of percolating filters than on those of activated sludge plants. The effect of the decline in the temperature of sewage during the winter is to reduce the activity of the macroinvertebrates. The growth rate of the film may also be reduced with temperature but generally to a smaller extent, and a net accumulation of film usually results. This may cause partial or complete blocking of the medium. Fungal growths flourish and at the same time worms and fly larvae, which would be present in the surface layers during the summer, retreat into the depths of the bed. The unequal effect of low temperature on film growth and animal activity limits the rate at which sewage can be applied to a filter, if an effluent of good quality is to be achieved. The arrival of warmer weather during the spring gives rise to a fairly rapid depletion of accumulated film in the filter and a corresponding increase in the concentration of suspended matter in the effluent. This phenomenon has not been completely explained. It has been attributed to the increased activity of the fly and worm populations in response to the temperature and also to microbial lysis within the film. The findings (Bruce, 1971) of pilot plant studies on high-rate filters at the Water Research Centre, Stevenage, showed a very marked sensitivity to temperature, the minimum winter BOD removal rate being less than half the maximum summer BOD removal rate.

13.5   PROCESS DESIGN

Aerobic biofilters are categorized as low-rate or high-rate, depending on the applied hydraulic and organic loading rates. Loading parameter ranges for both categories are summarized in Table 13.1. The conventional low-rate filter, used in domestic wastewater treatment, is typically loaded at about 0.1 kg BOD_5 m^{-3} d^{-1}, the corresponding hydraulic loading being about 1 m^3 m^{-2} d^{-1}. Space requirements are thus considerably greater than required for the activated sludge process (see to Table 12.1), even compared with the extended aeration version of the latter.

Table 13.1

Parameter	Low-rate filter	High-rate filter
Hydraulic loading (m d^{-1})	1–4	10–40
Organic loading (kg BOD$_5$ m^{-3} d^{-1})	0.1–0.3	0.3–2.0
Depth (m)*	1.5–3.0	1.5–6.0
Recirculation	none	usual

* Stone and similar solid aggregate media filters are generally not more than 2 m in depth, while high-rate plastic medium filters may be up to 6 m deep (so-called biotowers).

In high-rate filters, the recirculation of filter effluent permits the use of higher organic and hydraulic loading rates. A schematic outline of recirculation options is shown in Figure 13.3.

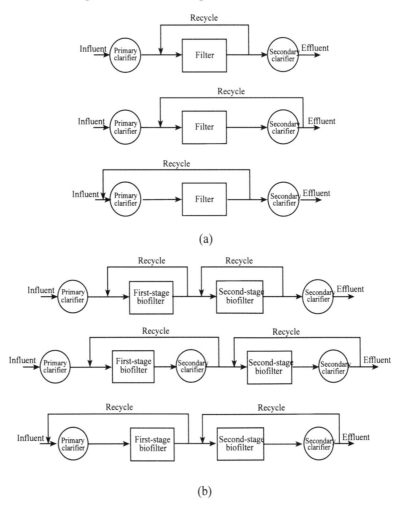

Figure 13.3
Recirculation options for high-rate biofilters: (a) single-stage filters; (b) two-stage filters
(Reproduced from Metcalf, 1991, with permission of the Macgraw-Hill Companies).

Because of the biological complexity of filter systems, accurate modelling of reactor kinetics has proved to be difficult. However, a number of empirical equations, quantifying filter performance, have been proposed (Metcalf and Eddy Inc, 1991). The NRC (1946) equations are amongst the simplest and most useful for design purposes. They were derived from an extensive study of the operating records of trickling filter plants serving military installations in the United States. The formulae are applicable to single-stage and multi-stage stone media systems, with varying recirculation rates. For a single-stage filter:

$$E_1 = \frac{1}{1 + 0.44\sqrt{\dfrac{W}{VF}}} \qquad (13.1)$$

where E_1 is the fractional efficiency of BOD removal for the biofilter, inclusive of recirculation and sedimentation; W is the BOD_5 loading rate (kg d^{-1}); V is the filter volume (m^3), and F is a recirculation factor, calculated as follows:

$$F = \frac{1 + R}{(1 + R/10)^2} \qquad (13.2)$$

where $R = Q_r/Q$; F is the average number of passes through the filter. The term $R/10$ takes into account the experimental observation that the removal of organics appears to decrease as the number of passes increases. For the two-stage filter (see Figure 13.3), the suggested NRC equation is:

$$E_2 = \frac{1}{1 + \dfrac{0.44}{(1 - E_1)}\sqrt{\dfrac{W^1}{VF}}} \qquad (13.3)$$

where E_2 is the fractional efficiency of BOD_5 removal for the second stage, including recirculation and sedimentation; W^1 is the second-stage BOD_5 loading rate, kg d^{-1}.

The influence of temperature on high-rate filter performance (see the earlier comments on seasonal influences) should be borne in mind when applying the NRC formulae to local conditions. It is probable that the efficiency levels computed from the NRC equations would only be attainable under summer conditions in temperate climates.

Alternating double filtration (ADF) is a further variation in filter layout. It consists of two filters operating in series, each with its own sedimentation tank. At intervals from daily to weekly, the order of the filters is reversed. The relegation of one of the filters to a secondary role causes a depletion in accumulated film and hence controls clogging. Thus, while a single low-rate filter might be subject to excessive film growth if the applied BOD_5 concentration exceeds 500 mg l^{-1}, the ADF arrangement overcomes this difficulty.

13.6 DESIGN OF PHYSICAL FACILITIES

The main elements of an aerobic biofilter system are: (1) an influent feed system by gravity or pump; (2) an influent irrigation system; (3) containment for the filter medium; (4) provision for adequate ventilation of the filter medium; and (5) secondary sedimentation and recirculation of effluent, where required.

13.6.1 Influent feed

For a single-stage filter process without recirculatiion, the available site gradient will determine if influent pumping is necessary. The head loss through a filter bed may be taken as the bed depth plus an allowance for upstream and downstream hydraulic losses of, say, 0.5 m. Where such a natural ground fall is not available, the clarified effluent from primary sedimentation is discharged to a pump sump, from where it is pumped to the filter distribution system. Where recirculation is being applied, recycled effluent is also discharged to this sump.

13.6.2 Distribution systems

The standard method of clarified effluent irrigation on circular filters is by rotary distributor, which consists of two or more arms, mounted on a pivot at the centre of the filter and revolving in a horizontal plane, as shown in Figure 13.1. Each arm is designed as a distribution manifold with discharge points so spaced as to secure a uniform irrigation of clarified wastewater over the entire bed area (requires a decreasing discharge nozzle spacing with distance from the centre of rotation). The distributor may be driven by the jet reaction of the nozzle discharges or by a geared motor. The speed of rotation is selected to give an application interval of 4–12 mins, which has been found (Tomlinson and Hall, 1955) to give satisfactory process performance. A clearance of at least 0.15 m is usually allowed between the underside of the distribution arm and the top of the bed. Motor-driven distributors have the advantage of providing a constant rotational speed, while the speed of reaction-driven distributors is determined by the flow rate. In small-scale units of the latter type, the inflow rate may be syphon-controlled to secure a measure of uniformity in application rate and hence in the distributor rotational speed.

Rectangular filters are less frequently used. Wastewater irrigation systems for rectangular filters generally incorporate a travelling manifold, which traverses the filter length in both directions. A fixed manifold distribution system is normally used for wastewater irrigation on deep-bed filters, so-called biotowers. These units incorporate self-sup-

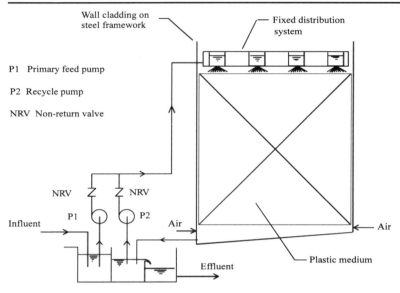

Figure 13.4
Typical Biotower
biofilter general
layout

porting plastic media and may be up to 6 m deep, with a correspondingly reduced plan area. Biotower filters are typically used at high
hydraulic and organic loading rates to achieve a 40–70% BOD reduction. A typical biotower filter layout is shown in Figure 13.4.

13.6.3 Filter structure

Stone or similar mineral medium filters are usually not more than 2 m
deep and are contained within reinforced concrete walled enclosures.
The floor is graded to provide under-drainage to a central or peripheral
collector channel.

Biotower filters, incorporating self-supporting plastic media, are generally protected by a sheeting-clad steel framework which absorbs wind
and other forces.

13.6.4 Ventilation

The ventilation of filters takes place by convective air movement
through the medium due to a temperature-related density difference
between the filter air and the ambient air. It is therefore important to
provide adequate air access openings at the base of the medium to
facilitate natural ventilation. Such ventilation is conveniently provided
by designing the underdrain collector system to flow partly filled, so
that it also acts as an air distribution system, and by ensuring that it
is adequately vented. A total vent area of 0.4% of the plan area of the
filter has been proposed (Metcalf and Eddy, Inc. 1991) as a design
guide.

13.6.5   Filter media

The essential requirements for a filter medium are that it should be inert, be of sound mechanical strength, and possess within its bulk an extensive area of exposed surfaces over which the liquid to be treated can be passed. Adequate void spaces must exist between adjacent surfaces to allow for some accumulation of biological film, for free passage of liquid and suspended matter, and for access of air. Performance can be correlated with the specific surface of the medium, provided that the degree of film accumulation is not such as to cause clogging. Thus with low film conditions, a 25 mm medium of a given type will produce a better effluent than a 50 mm medium of the same type. Also, for a given size of medium, rough-surfaced materials such as clinker have a marginally superior performance to that of smooth materials such as gravels. In practice, a compromise between the conflicting requirements for large specific surface and adequate void space results in the use of a medium of about 50 mm size in natural aggregate or crushed rock material. Specifications for filter media can be found in BS1438: 1971 and ASCE Manual 13.

A variety of lightweight plastic filter media has been developed. Some representative types are compared with stone media in Table 13.2.

Table 13.2 Filter media characteristics

Medium type	Size/arrangement	Specific surface $(m^2\ m^{-3})$	Voids (%)
Blast-furnace slag	80–130 mm, random packing	40	50
Smooth rock (basalt)		40	53
Plastic 1	PVC foil modules	85	98
Plastic 2	polystyrene sheets, close-packed	187	94
Plastic 3	separate PVC tubes, continuous vertical lengths	220	94
Plastic 4	Polystyrene sheets	82	94

Note: Plastic media 1 to 4 are representative of available proprietary systems.

The large voidage and predominantly vertical orientation of the surfaces of plastic media, which discourages the development of thick film layers, permit unrestricted flow of liquid over the medium surface. It is probable that macroinvertebrates play only a minor role in plastic media filters, which are continuously wetted at a high rate from a fixed manifold distribution system.

13.7   COMPARISON OF ACTIVATED SLUDGE AND BIOFILTER PROCESSES

The main advantages of biofilter processes over activated sludge (AS) systems are: (1) Because it is an attached film process, the biofilter is not susceptible to the 'sludge bulking' phenomenon, which can greatly reduce the efficiency of the AS process. (2) In its low-rate form, the aerobic biofilter uses less energy than the AS process.

The main disadvantages are: (1) To achieve the same degree of BOD removal, aerobic biofilters require a much larger process volume than the AS process; pre-sedimentation is also required for wastewaters containing settleable solids. Hence the capital cost of biofilter systems may be significantly higher than that of AS systems. (2) Biofilter systems are operationally less flexible than AS systems; biofilter systems suffer a greater reduction in performance at low temperature than do AS systems. (3) Under certain conditions, biofilter systems may give rise to odour nuisance (high-rate systems) and fly nuisance (low-rate systems).

13.8   NITRIFICATION

As noted earlier, reduced nitrogen species are converted to nitrite, which is further converted to nitrate, in low-rate biofilters. The variation in ammonia concentration with distance from the surface of a low rate biofilter treating domestic wastewater is shown in Figure 13.2. The extent of nitrification in biofilters is very much influenced by

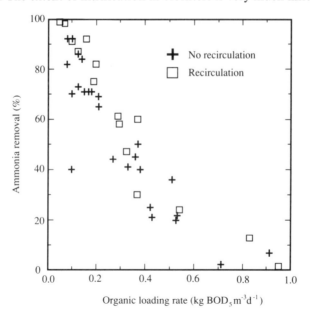

Figure 13.5
Influence of organic loading on nitrification in rock-filled biofilters. (Reprinted from USEPA, 1993 with permission of Technomic Publishing Company, Inc.)

the organic space loading rate, as shown in Figure 13.5. The degree of nitrification is seen to decrease approximately linearly with increasing organic loading rate, reducing from in excess of 90% at organic loading rates less than 0.1 kg BOD_5 m^{-3} d^{-1} to about 40% at an organic loading rate of 0.4 kg BOD_5 m^{-3} d^{-1}. It would appear that very little nitrification takes place at organic loading above 0.6 kg BOD_5 m^{-3} d^{-1}.

13.9 SLUDGE PRODUCTION IN BIOFILTERS

There is not much published data on the rate of sludge production in aerobic biofiltration processes. The rate of humus sludge production in low-rate pilot plant biofilters has been reported (Bruce and Boon, 1970) to vary within the range 0.08–0.5 kg per kg BOD_5 removed, the higher value relating to spring time, when there is a marked shedding of filter slime and the lower value relating to summer time. The corresponding mean rate of film discharge was found to be 0.22 kg per kg BOD_5 removed, which agrees reasonably well with observed activated sludge production rates in extended aeration systems (see to Chapter 12). The average rate of sludge discharge from high-rate pilot plants, treating settled domestic sewage, has been found (Bruce and Boon, 1970) to be in the range 0.63–1.0 kg per kg BOD_5 removed, with little seasonal variation in this range.

13.10 ROTATING BIOLOGICAL CONTACTOR

The rotating biological contactor (RBC) is an attached biofilm system which typically consists of a series of circular discs (biodisk) mounted on a horizontal shaft placed in a semicircular trough with approximately 40–45% disk submergence. Alternatively, a cylindrical mesh drum filled with random packing (biodrum) may be used. The biodisk or biodrum is rotated at a speed that allows adequate attached biofilm development. Oxygen transfer is achieved by the exposure and renewal of air–water interfaces as the contactor rotates and the wastewater lifted out by the rotating device trickles back down into the sump. This cyclic immersion of the biofilm also provides the opportunity for the adsorption and uptake of organics from the wastewater. Although the concept was first put forward by Weigand in 1900 and subsequently tested by Doman in 1929, the process was not commercially developed as a wastewater treatment process until the 1960s (Huang and Bates, 1980).

While many types of proprietary RBC systems have been developed, mainly for small-scale application (< 500 PE), they generally conform to the following design parameter ranges:

disc or drum diameter: 0.7–3.5 m
rotational speed: 0.5–10 rpm
biofilm loading: 3.0–6.0 g BOD_5 m^{-2} d^{-1}

As in conventional biofiltration processes, the RBC process requires pre-treatment by primary sedimentation and post-treatment by secondary sedimentation. Experimental biodisk studies (Bruce and Merkens, 1975), treating settled domestic sewage, have shown that where the biofilm loading is less than about 6 g BOD_5 m^2 d^{-1}, a settled effluent BOD_5 less than 20 mg l^{-1} can be obtained. The trough volume is determined by the design biofilm loading rate and the strength of the settled wastewater. It has been found advantageous (Pike *et al.*, 1982) to subdivide the trough into a set of chambers in series, thus simulating a plug flow regime and hence increasing the average conversion rate for the processs unit.

RBCs are vulnerable to freeze-up and subsequent mechanical damage in cold weather and hence are normally covered in climates subject to freezing conditions.

REFERENCES

Bruce, A. M. (1969) *Process Biochem.*, **4**, No.4, 19.
Bruce, A. M. (1971) *Proc. 5th. Intern. Conf. Wat. Pollut. Res.*, Pergamon Press, Oxford.
Bruce, A. M. and Boon, A. G. (1970) *Aspects of High-rate Biological Treatment of Domestic and Industrial Wastewaters*, Public Works Municipal Services Congress, WRC, Stevenage, reprint no. 604.
Bruce, A. M. and Merkens, J. C. (1975) *Proc. 8th. Public Health Engineering Conference*, Loughborough University of Technology, Loughborough, UK.
Huang, Ju-Chang and Bates, V. T. (1980) *J. Water Pollut. Control Fed.*, **52**, No. 11, 2686–2703.
Metcalf and Eddy, Inc. (1991) *Wastewater Engineering*, 3rd edn., McGraw-Hill Publishing Co., New York.
National Research Council (1946) *Sewage Works J.*, **18**, No. 5.
Pike, E. B., Carlton-Smith, C. H., Evans, R. H. and Harrington, D. W. (1982) *J. Water Plllut. Control*, **81**, 10–27.
Tomlinson, T. G. and Jenkins, S. H. (1947) *J. Inst. Sewage Purif.*, **Pt. 2**, 94.
Tomlinson, T. G. and Hall, H. (1955) *J. Inst. Sewage. Purif.*, **Pt. 1**, 40.
USEPA (1993) *Nitrogen Control*, Technomic Publishing Co., Inc., Lancaster, PA, USA.

Related reading

Cooper, P. F. and Atkinson, B. (!981) *Biological Fluidised Bed Treatment of Water and Wastewater*, Ellis Horwood Ltd, Chichester, UK.
Qasim, S. R. (1994) *Wastewater Treatment Plants, Planning, Design and Operation*, Technomic Publishing Company, Lancaster, PA, USA.
USEPA (1993) *Municipal Wastewater Treatment Technology*, Noyes Data Corporation, New Jersey, USA.

14

Gas-Liquid Transfer

14.1 GAS TRANSFER

By gas transfer is meant the mass movement of gas to or from solution, which occurs when a liquid surface is in contact with a gas phase with which it is not in equilibrium. In chemical engineering practice the uptake of gas by a liquid is generally termed 'absorption' while the removal of gas from a liquid is usually termed 'desorption' or 'stripping'.

In water and wastewater engineering practice the gas phase is usually air and hence gas transfer is generally termed 'aeration'. The gases of interest to the water and wastewater engineer are O_2, CO_2, CH_4, H_2S, NH_3 and Cl_2. The driving force that activates the transfer of these gases to and from water is the difference between their concentration in solution and their solubility under the prevailing conditions.

14.2 GAS SOLUBILITY

The main factors that influence gas solubility in water are: the water temperature, the partial pressure of the gas in the gas phase, the dissolved solids concentration in the water phase and the chemical nature of the gas.

The solubility of gases, unlike that of solids, decreases with increasing temperature. For partial pressures up to 1 atm and for gases which do not react with water to any great extent, the equilibrium concentration of gas in solution at a given temperature is proportional to the partial pressure of the gas in contact with the water in accordance with Henry's Law:

$$c_s = Hp \qquad (14.1)$$

where $c_s(mgl^{-1})$ is the saturation or equilibrium concentration of the gas in solution, p (atm) is the partial pressure of the gas in the gas phase in contact with the water and H is the solubility coefficient. Henry's Law applies to most of the gases of interest in water and wastewater engineering including O_2, CH_4, CO_2, H_2S. The latter two gases undergo reaction in solution.

Dissolved CO_2 reacts with water as follows:

$$CO_2 + H_2O \Leftrightarrow H_2CO_3 \tag{14.2}$$

$$H_2CO_3 \Leftrightarrow H^+ + HCO_3^- \tag{14.3}$$

$$HCO_3^- \Leftrightarrow H^+ + CO_3^{2-} \tag{14.4}$$

For the conditions normally encountered in water engineering the concentration of H_2CO_3 will not exceed 1% of the CO_2 concentration, while the dissociation constants for reactions (14.3) and (14.4) are also very small (see p. 9). Hydrogen sulphide reacts in solution as follows:

$$H_2S \Leftrightarrow H^+ + HS^- \tag{14.5}$$

$$HS^- \Leftrightarrow H^+ + S^{2-} \tag{14.6}$$

It is apparent from equations (14.5) and (14.6) that the dissolved form of H_2S depends on the pH of the solution.

Ammonia (NH_3) and chlorine (Cl_2) are highly soluble gases which readily react with water. Their pressure–solubility relationships deviate from Henry's Law.

Water solubility data for the above gases are presented in Tables 1.7, 1.8 and 1.9 in Chapter 1 and may also be accessed through the computer program GASSOL. It is clear from the tabulated data that the gases that react with water, including CO_2, H_2S, NH_3 and Cl_2, are considerably more soluble than the remaining non-reactive gases.

14.3   MECHANISM AND RATE OF GAS TRANSFER

When a water surface is exposed to a poorly soluble gas with which it is not already in equilibrium, it is assumed that the water at the interface becomes instantaneously saturated with the gas and that the gas is transported into the body of the liquid by the process of molecular diffusion:

$$\frac{\partial m}{\partial t} = -D \frac{\partial c}{\partial x} \tag{14.7}$$

where $\partial m/\partial t$ is the rate of gas transport across unit area of interface, D is the coefficient of molecular diffusion and $\partial c/\partial x$ is the concentration gradient normal to the interface. The simplest physical model of inter-

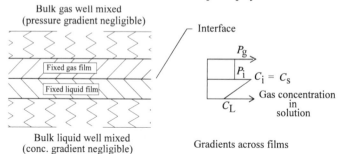

Figure 14.1
Two-film gas
transfer model

face conditions is that proposed by Lewis and Whitman (1924) as shown in Figure 14.1. They assumed that resistance to gas transfer resided in fixed gas and liquid films at the gas–liquid interface. The movement of gas across the gas film implies the existence of a pressure gradient in the gas film and hence the gas pressure P_i at the interface is less than the bulk gas pressure P_g. However, for gases of low solubility the rate of transfer of gas to the liquid phase is slow and hence the pressure gradient in the gas phase is negligibly small. Whitman and Lewis assumed a linear concentration gradient across the liquid film:

$$\frac{\partial c}{\partial x} = \frac{C_L - C_s}{h} \tag{14.8}$$

where C_s is the saturation or interface concentration of gas in solution, and C_L is the gas concentration in the bulk liquid (assumed to be well-mixed). The rate of gas transfer across a gas–liquid interface of area A is calculated from equations (14.7) and (14.8) to be

$$A\frac{\partial m}{\partial t} = -\frac{AD}{h}(C_L - C_s) \tag{14.9}$$

Dividing by the liquid volume V:

$$\frac{A}{V}\frac{\partial m}{\partial t} = -\frac{AD}{Vh}(C_L - C_s)$$

Hence

$$\frac{dc}{dt} = -a\frac{D}{h}(C_L - C_s) \tag{14.10}$$

or

$$\frac{dc}{dt} = -k_L a(C_L - C_s) \tag{14.11}$$

where $a = A/V$, k_L is called the 'liquid film coefficient' (also called the 'absorption coefficient') and the product $k_L a$ is the 'overall transfer coefficient'. Although it can be verified experimentally that an equation such as (14.11) accurately represents gas-transfer kinetics in systems of the O_2/water and CO_2/water types, it is unlikely that the simple liquid film model, on which the equation is based, correctly represents actual conditions at the interface. In the aeration systems used in water and wastewater treatment processes, large areas of air–water interface are created and continually renewed. When an element of interface is created, the initial rate of gas transfer to or from a non-equilibrated liquid is high, but is reduced as the concentration gradient reduces (equation 14.7). Thus the more rapidly an element of interface is renewed the higher will be the mean rate of gas transfer through it. Also, since gas concentration gradients in the bulk water are always very small, indicating little resistance to gas transport within the bulk liquid phase, it is apparent that resistance to gas transfer resides in the organized water at the interface. It is probable that the effective thickness of this high-resistance film is influenced by the level of turbulence in the bulk liquid. Thus the overall transfer coefficient, which is a constant for a given aeration system, is dependent on the total area of air–water interface created, the rate of interface renewal, the level of turbulence in the liquid and temperature.

14.4 OXYGEN TRANSFER

The main function of aeration in wastewater treatment is the solution of oxygen in the mixed liquor of the activated sludge process. In the activated sludge process, as in other similar biochemical processes, the rate of oxygen input to the system must balance the rate of oxygen uptake, since, owing to its low solubility, the storage of oxygen in the system is negligible. Aerators are also sometimes used in polluted rivers and lakes to prevent excessive oxygen depletion.

Aerators for oxygen transfer are rated in terms of oxygenation capacity (OC), which is defined as the rate at which an aerator transfers oxygen to clean water under standard conditions, i.e. at a temperature of 20°C (or 10°C), an atmospheric pressure of 1 atm and a zero concentration of oxygen in solution. Under these conditions equation (14.11) becomes:

$$\frac{dc}{dt} = k_L a C_{*(20)} \tag{14.12}$$

where $C_{*(20)}$ is the equilibrium or saturation concentration of oxygen in the water being aerated, at a water temperature of 20°C. It should be noted that $C_{*(20)}$ may differ significantly from $C_{s(20)}$, the saturation concentration for a water in equilibrium with atmospheric air. This is particularly the case for dispersed air systems with submerged diffusers, where $C_{*(20)}$ typically exceeds $C_{s(20)}$. With surface aeration systems, $C_{*(20)}$ may be taken as equal to $C_{s(20)}$ for practical computational purposes. Since

$$OC = V \frac{dc}{dt} \tag{14.13}$$

then

$$OC = k_L a C_{*(20)} V \tag{14.14}$$

where V is the liquid volume.

The conditions under which aerators operate in the activated sludge process are quite different from the standard conditions to which their OC-values relate. The more significant differences are the presence of solids in suspension, the presence of surfactants in solution and the temperature difference. These factors may influence both the value of $k_L a$ and C_s. Their effects on these parameters are taken into account by the introduction of the empirical coefficients α and β, which are defined as follows:

$$\alpha = \frac{k_La \text{ in suspension}}{k_La \text{ in tapwater}} \qquad (14.15)$$

$$\beta = \frac{C_s \text{ for suspension}}{C_s \text{ for tapwater}} \qquad (14.16)$$

The concentration of suspended solids in the mixed liquor of activated sludge plants is usually within the range 1000–6000 mg l^{-1}. Since suspended solids may alter the bulk and surface properties of the suspending medium, they are likely to affect gas transfer kinetics (Karmo, 1972). Suspended solids effect an increase viscosity which tends to reduce the level of turbulence and rate of surface renewal. Particles that have a preferential affinity to adsorb on the water surface may influence surface hydrodynamics. On the credit side, air bubbles may become attached to the solids in suspension and be carried down into the bulk liquid.

Surface-active agents are present in most wastewaters treated by the activated sludge process and for this reason their influence on oxygen transfer has been widely studied. By reducing surface tension, they effect a increase in the diameter of spherical bubbles and a corresponding increase in the air–water interfacial area per unit volume of air. On the negative side, however, they tend to attach to the bubble surface, where they constitute a barrier to oxygen transfer. The latter effect is more marked in diffused air systems, where the rising bubbles have a sufficiently long residence time for this layer to become established.

Test results show that the operational α-value in activated sludge processes is typically less than unity. It is well established (Barnhart, 1969) that surface-active agents in solution significantly reduce the oxygen transfer rate of diffused air aeration systems, resulting in a typical operational α-value range of 0.4–1.0. The corresponding typical α-value range for surface aerators is 0.8–1.2. It should be noted that an α-value in excess of unity infers an enhanced rate of oxygen transfer under process conditions. This may arise where there is a rapid renewal of the air–water interface and hence insufficient time for the formation of a surfactant interfacial barrier to be formed.

The value of the β-factor under process conditions is also typically less than unity and clearly depends on the dissolved solids concentration of the aqueous suspension being aerated. In activated sludge processes, the value of β is generally between 0.9 and 1.0.

14.5 INFLUENCE OF TEMPERATURE ON OXYGEN TRANSFER

The oxygen absorption coefficient k_L increases with increasing temperature, due mainly to the temperature-dependent nature of those water properties that affect gas transfer, namely diffusivity, surface tension and viscosity. Oxygen diffusivity in water increases with increasing temperture. Surface tension and viscosity, both of which affect the surface renewal rate, decrease with increasing temperature. These latter properties also influence the energy required to create air–water interfaces. The total area of air–water interface created per unit energy input should increase with increasing temperature, as surface tension and viscosity correspondingly decrease. This infers that the overall transfer coefficient k_L should be affected to a greater extent by temperature than is the absorption coefficient k_L.

In rating aeration systems, it is usual practice to relate performance to a standard temperature of 20°C. The value at any temperature may be related to its value at 20°C by a temperature coefficient f, as follows:

$$(k_L a)_{20} = (k_L a)_T \, f^{(20-T)} \tag{14.17}$$

where T denotes the temperature (°C) at which k_{La} is measured. Barnhart (1969) has reviewed studies on the temperature dependence of oxygen transfer in diffused air aeration systems, for which he reported an average value of 1.02 for f. In surface aeration systems, the reported value range for f is 1.012–1.047.

It is also worthy of note that the rate of oxygen transfer by an aeration system, at zero oxygen concentration and within the temperature range generally encountered in water and wastewater engineering practice, is not greatly affected by temperature since the increase in k_{La} with rising temperature is offset by a corresponding decrease in the solubility of oxygen. Using the oxygen solubility values given in Table 1.7 and an f value of 1.02, calculations show that, while there is an increase of 48.6% in the value of $k_L a$ due to a temperature rise from 0°C to 20°C, the actual rate of oxygen transfer, at zero oxygen concentration, by an aeration system at 20°C is less than its rate of transfer at 0°C, but only by about 8%.

14.6 AERATION SYSTEMS USED IN THE ACTIVATED SLUDGE PROCESS

The three main types of aeration system used in the activated sludge (AS) process are the following:

(a) dispersed air systems;

(b) mechanical surface aeration systems; and

(c) combinations of (a) and (b).

14.6.1 Dispersed air systems

In dispersed air systems, air is introduced to the mixed liquor through 'diffusers', located at some distance below the surface. Dispersed air systems are broadly classified as coarse or fine bubble systems depending on the size of bubble generated. Although the boundary between these two categories is not very sharply defined, fine bubble diffusers typically produce bubbles in the 2–5 mm size range in clean water, while coarse bubble diffusers produce bubbles in the size range 6–10 mm.

Figure 14.2 Diffused air activated sludge process. (Reproduced courtesy of Bowen Water Technology, Kilkenny, Ireland)

Coarse bubble diffusers typically consist of drilled holes or slots in a submerged air distribution system. The bubbles produced may be smaller than the orifice size, through being sheared off by the water pressure. The oxygen transfer efficiency (*OTE*) of coarse bubble systems is generally much lower than fine bubble systems (*OTE* is defined as the percent of oxygen transferred per metre submergence, at zero dissolved oxygen concentration).

Fine bubble diffusers (USEPA, 1989) are constructed from a range of materials including ceramics, porous plastics and perforated membranes. Ceramics and porous plastics have interconnected pore structures through which the air flows to be discharged through the top surface as a bubble stream. Their pore structure is rigid and hence subject to fouling either due to impurities in the air or precipitation from the water side. To prevent clogging of porous diffusers, it has been found necessary (Pasveer and Sweeris, 1965) to reduce the dust content of the air used to a level not greater than $0.036 \, \text{mg m}^{-3}$. This is accomplished by filtration.

The perforated membranes are made from either thermoplastic or elastomeric sheet materials (plasticized PVC, EPDM rubber, neoprene rubber) of thickness 1–2 mm. They are mechanically perforated with a pattern of small holes or slits, which expand as the airflow increases, hence they are commonly known as flexible membrane diffusers.

Oxygen transfer from an individual air bubble may be considered to take place in three consecutive stages: bubble formation, bubble ascent and bubble escape at the surface. The oxygen transfer rate is high during the bubble formation phase, since a new air–water interface area is being created and the interfacial oxygen concentration gradient is therefore high. The extent of oxygen transfer during ascent depends on the rate of interface renewal, bubble surface area and bubble contact time, factors that are dependent on bubble size and subsequent depth of the diffuser. Barnhart (1969) found that for given volumes of air and water, the overall transfer coefficient had a maximum value at a bubble size of 2.2 mm. Downing (1960) found that the average value of the overall transfer coefficient decreased with increasing depth of the diffuser. The total contact time, which effectively equals the time of ascent, is greatly reduced if the water, into which the bubble is released, has itself an upward velocity as is the case in spiral flow tanks, where diffusers are located on one side of the tank only. The final phase of oxygen transfer is due to surface disturbance generated by bubble escape. Experimental evidence suggests that in most cases this is the least significant of the three phases.

Many attempts have been made to correlate absorption coefficients for bubble aeration systems. However, because of the many variables involved and the strong influence on test results of the hydrodynamic conditions obtaining in the test vessels, correlation is often not feasible and is probably only meaningful in tests carried out on single bubbles. Reported mean values of k_L vary from 3.67 mm min^{-1} for a single large bubble in a narrow tube (Adeney and Becker, 1918, 1919, 1920) to 41.67 mm min^{-1} for bubbles in a highly turbulent aeration tank (Holroyd, 1952).

The oxygenation capacity of a bubble aeration system is most reliably obtained by an *in situ* experimental determination, as later described.

The schematic outlines of a typical diffused air aeration system is shown in Figure 14.3.

14.6.2 Mechanical Surface Aeration Systems

Mechanical surface aeration systems effect oxygen transfer by creating and rapidly renewing large areas of the air–water interface. Such interfaces result from a combination of entrainment of air, spraying of water into the air and the generation of a high level of surface turbulence.

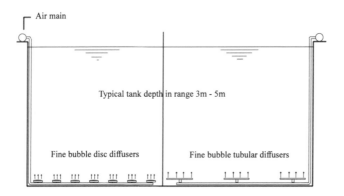

Figure 14.3
Diffused air aeration system layout

Figure 14.4 Activated sludge process using a vertical shaft surface aerator

Mechanical surface aerators may be broadly classified in the following categories:

(a) Impeller-type devices mounted on a vertical shaft.

(b) Rotor-type devices, mounted on a horizontal shaft.

Vertical shaft devices are either of the plate or turbine type.

Plate aerators (Downing *et al.*, 1960) consist of a horizontal circular plate, mounted on a vertical shaft, to the underside of which is attached a set of vertical radial or curved blades. In operation, the top of the plate is located at or slightly below the still water surface level. The rotation of the disc causes liquid to be discharged radially, creating a circular hydraulic jump in which air is entrained. Air may also be sucked down through openings in the plate to the low-pressure zones created behind each vane as the plate rotates. A plate aerator on its own may not provide adequate mixing in an aeration tank. Supplemental mixing can be provided by extending the plate shaft downwards and attaching a mixing impeller at a lower level.

Turbine aerators (Downing *et al.*, 1960) effect oxygen transfer primarily by a pumping action. The impeller, which is located at the liquid surface, is designed to pump large quantities of liquid at a low head. Because the impeller is located at the liquid surface a large amount of air is entrained in the pumped liquid, which is thrown upwards and outwards in a low trajectory, creating considerable turbulence as it strikes the liquid surface. In deep tanks, turbine

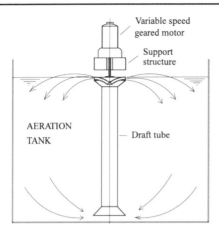

Figure 14.4a
Typical
arrangement of a
vertical shaft
surface aeration
system

aerators may be used in conjunction with a draft tube, which extends downwards to within a short distance of the tank bottom, as shown in Figure 14.4a. This arrangement ensures good mixing of the tank contents.

Rotor-type or 'brush' aerators consist of a horizontal revolving shaft, to which are attached flat or angular blades, projecting radially outwards. The overall diameter of the brush rotors is usually between 0.6 m and 1.0 m; their immersion depths may vary between 100 mm and 300 mm, depending on diameter.

Brush aerators (Casey, 1971) effect oxygen transfer by entraining air in the liquid spray which is thrown upwards and outwards by the revolving blades. Considerable surface turbulence is also generated. Brush aerators are generally used in closed- loop aeration tanks of the oxidation ditch type, as shown in Figure 12.9.

It should be noted that aeration systems are required to provide adequate mixing as well as transfer the required amount of oxygen to solution. Hence, the geometry of the aeration tank must be matched to the mixing characteristics of the aeration device. In some situations it may be necessary to supplement a diffused air system by a mechanical mixing device to ensure an adequate distribution of mixing throughout the aeration tank.

14.7 EXPERIMENTAL DETERMINATION OF OXYGENATION CAPACITY (ASCE, 1984)

The field OC of an aeration system is the rate at which it would transfer oxygen to clean water at a temperature of 20°C, a zero concentration of oxygen in solution and a prevailing atmospheric pressure of 1 atm. In general, the prevailing test conditions will not be standard and hence temperature and pressure corrections have to be applied to the test results. Combining equations (14.14) and (14.17), the resultant expression for OC becomes:

$$OC = k_L a_{(T)} f^{(20-T)} C_{*(20)} V \frac{P_s}{P_b} \qquad (14.18)$$

where $C_{*(20)}$ is the saturation or equilibrium concentration of oxygen in solution (kg m^{-3}) under test conditions at a water temperature of 20°C, V is the aerated water volume (m^3), P_b is the prevailing atmospheric pressure (atm) at the test site and P_s is the standard barometric pressure of 1 atm.

For surface aeration systems, the value of $C_{*(20)}$ can be taken as being equal to $C_{s(20)}$, the saturation concentration corresponding to an air pressure of 1 atm, which, as given in Table 1.7, is 9.08 mg l^{-1}. For diffused air systems, however, the value of $C_{*(20)}$ exceeds $C_{s(20)}$ by an amount which depends on the diffuser submergence and, hence, has to be determined experimentally.

As shown in section 14.3, the rate of oxygen transfer during an aeration process is represented by the expression:

$$\frac{dc}{dt} = k_L a(C_* - C)$$

or

$$\frac{dc}{(C_* - C)} = k_L a(dt)$$

Integrating this equation between an initial concentration of oxygen in solution of C_i at time zero and a concentration C_t after an aeration interval of t:

$$\int_{C_i}^{C_t} \frac{dc}{(C_* - C_L)} = \int_0^t k_L a(dt)$$

This gives

$$\ln\left[\frac{(C_* - C_i)}{(C_* - C_t)}\right] = k_L a \cdot t \tag{14.19}$$

or

$$\ln(C_* - C_t) = \ln(C_* - C_i) - k_L a \cdot t \tag{14.20}$$

If the natural log of oxygen deficit $(C_* - C_t)$ is plotted as a function of aeration time t, then a straight-line graph is obtained in accordance with equation (14.20), as illustrated in Figure 14.5. The overall transfer coefficient $k_L a$ is the slope of this line and has the units time^{-1}.

The *unsteady-state* re-aeration method is the usual experimental method for k_L determination. The water is first de-oxygenated by gas stripping or by the addition of a chemical reducing agent such as sodium sulphite (Na_2SO_3). To accelerate the chemical reduction of oxygen, a catalyst such as Co^{2+} is added at a concentration of about 0.1 mg l^{-1}. Because it is necessary to leave the aeration system in operation during the deoxygenation process, the amount of sulphite added is usually in excess of the stoichiometric amount by 10–20%. The rate of re-oxygenation is then monitored, usually by a number of DO meters located at

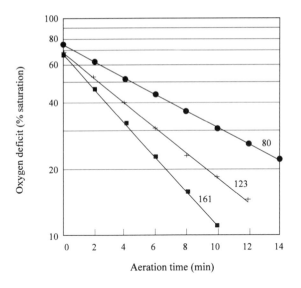

Figure 14.5
Results of an oxygenation capacity test on a bladed rotor. Rotor dia. 700 mm; figures on graphs are immersion depths in mm

different depths. For a reliable determination of $k_L a$, it is desirable that the re-oxygenation measurement should extend over a wide value range, e.g. 10%–90% saturation. Since aeration tanks are normally well mixed, the spatial variation in oxygen concentration at a given instant in time tends to be small. The measured re-oxygenation rate data are used to compute the value of $k_L a$ for the system, using the semi-log graphical method or the three-parameter fit method as described in the following paragraphs.

The simplest method of $k_L a$ determination is the semi-log graphical method, in which the saturation deficit is plotted on semi-log paper as a function of time, resulting in straight-line plots, as shown on Figure 14.5. The value of $k_L a$ is evaluated as the slope of a semi-log plot and is corrected for temperature in accordance with equation (14.17). The OC-value for the aeration system is computed by insertion of the appropriate parameter values in equation (14.18). This method is suitable for surface aeration systems where the saturation concentration can be assumed to be the same as the concentration in equilibrium with atmospheric air.

To illustrate the three-parameter fit method, equation (14.20) is written in the form:

$$C_t = C_* - (C_* - C_i)\exp(-k_L a \cdot t) \qquad (14.21)$$

Non-linear regression is used to find the best-fitting values of the three parameters C_*, C_i and $k_L a$. An example of the output from such a computation is presented in Figure 14.6. Details of the computational procedure, together with an example, are given in Appendix C (a listing of the computer program OCFIT is given in Appendix A). This is the preferred method of $k_L a$ computation as it is based on a

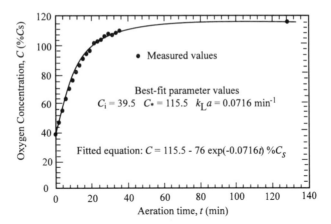

Figure 14.6
Computation of $k_L a$ using the three-parameter best-fit method. Diffused air system, diffuser submergence 4 m. %C_s is the reading on an air-calibrated DO meter

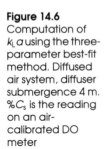

best-fit model of the experimental measurements. It has the advantage of not requiring an assumed or measured value for the saturation concentration C_*, which is of particular significance for diffused air systems where the saturation concentration is a function of the system configuration.

The steady-state oxygen transfer rate (OTR) by an activated sludge aeration system, operating at a fixed DO concentration C (mg l^{-1}) and at a temperature T (°C), is related to its standard OC-value as follows:

$$OTR = \alpha OC f^{(T-20)} \frac{(\beta C_{*(T)} - C)}{C_{*(20)}} \frac{P_b}{P_s}$$ (14.22)

The oxygen transfer rate in diffused air systems can be measured under field operating conditions by monitoring the oxygen concentration in the off-gas from the aeration system. This is conveniently done using an air-calibrated DO meter. This procedure can be used as a steady-state method for measuring oxygen transfer in activated sludge processes operating at a constant loading rate and hence at a fixed DO level.

14.8   ENERGY CONSUMPTION AND OXYGENATION EFFICIENCY

The standard aeration efficiency (SAE) of an aeration system is usually defined in terms of the energy expended per unit mass of oxygen dissolved under standard OC test conditions:

$$SAE = OC \text{ per power input } \text{kg } O_2 \text{ kWh}^{-1}$$ (14.23)

The SAE value may be based on net power consumption, which is exclusive of motor and gearbox efficiencies, or gross energy, i.e. wireto-water efficiency. The net SAE of aeration systems used in wastewater treatment practice typically varies in the range 1.5–6.0 kg O_2 kWh^{-1}. Efficiently designed fine bubble diffuser systems in new condition may be expected to perform towards the upper end of this range, while the performance of coarse bubble systems and less efficient surface aeration systems may be expected to fall near the lower end of the range. In general, the oxygen transfer performance of vertical shaft surface aeration systems increases with energy input per unit tank volume, i.e. the performance improves as the tank size is reduced.

The energy efficiency of aeration systems under field operating conditions generally fall within a much narrower band than that given for SAE above. This is due to the fact that the performance of diffused air systems is more adversely affected (reduced α and/or β-factors) by contaminants such as surfactants than is the performance of surface aeration systems. While fine bubble diffused air systems are generally more efficient than surface aeration systems, they have the disadvantage of being susceptible to clogging due to either impurities in the air or chemical precipitation on the water side. Both types of aeration system are widely used in wastewater treatment practice.

The standard oxygen transfer efficiency ($SOTE$) is the fraction of the oxygen in the compressed air flow in a diffused air system that is taken into solution under standard conditions:

$$SOTE = OC/(0.228 \, M_a) \qquad (14.24)$$

where M_a is the air mass flow rate with an oxygen mass fraction of 0.228.

The $SOTE$ of fine bubble diffusers is typically in the range 4–6% per metre submergence for diffuser submergences in the range 6–10 m (USEPA, 1989), decreasing with increasing diffuser air flow rate. The $SOTE$ of coarse bubble systems is typically about half that of fine bubble systems.

14.9   GAS STRIPPING

14.9.1   Removal of carbon dioxide

Carbon dioxide (CO_2) removal is required for those groundwaters that have become supersaturated with the gas (for solubility data refer to Table 1.7). CO_2 is normally added to a water or wastewater which requires 'recarbonation' following lime treatment.

The basic transfer principles already outlined for oxygen are also applicable to CO_2 and hence the same transfer technology may be used. For gas removal it is important to keep the partial pressure of the gas in the gas phase as low as possible—this can be achieved by good ventilation. In addition to dispersed air systems, spray nozzles, cascades and forced-draft multiple tray aerators are used for CO_2 removal.

14.9.2   Removal of hydrogen sulphide

Hydrogen sulphide (H_2S) is found in some groundwaters and in effluents from anaerobic treatment systems. Even at low concentrations it gives rise to odour and corrosion problems and has a high chlorine demand. Its removal by aeration is complicated by it dissociation in solution (see equations (14.5) and (14.6)). To control the extent of formation of HS^- and S^{2-} it may be necessary to reduce the pH by acid addition. Added oxygen may also react with H_2S to form free sulphur. While the latter reaction reduces the H_2S content it may lead to further problems in chlorination and distribution (growth of sulphur bacteria in the distribution system). Although the general aeration principles already outlined are valid, because of the additional complicating factors involved in H_2S removal, pilot plant studies should be made prior to plant design.

14.9.3   Removal of ammonia by aeration

Ammonia is produced in most wastewaters including domestic sewage, through the microbial degradation of organic compounds containing nitrogen. Since it is an important nutrient, its removal from effluents is often necessary to control eutrophication in receiving waters. This can be achieved biochemically by nitrification (conversion to NO_3^-) followed by subsequent denitrification to N_2, in which form it is no longer an available source of nitrogen for most organisms. Ammonia can also be removed by aeration. For efficient air-stripping a high pH is required to convert ammonium ion to NH_3:

$$NH_4^+ + OH^- \Leftrightarrow NH_4OH \Leftrightarrow NH_3 + H_2O \qquad (14.25)$$

Because ammonia is a highly soluble gas (see Table 1.8), large volumes of air are required to effect its removal. The efficiency of the process increases with increasing temperature owing to the resulting reduction in gas solubility.

14.10   GAS TRANSFER IN PACKED COLUMNS

Inter-phase gas transfer is achieved in packed columns by the creation and renewal of the gas–water interface area as water trickles down through a bed of discrete packings of the type shown on Figure 14.7. The downward cascading water is bought into close contact with the gas contained in the voids of the packing. Depending on the requirements of the application the gas phase may be maintained at a fixed pressure (as in air saturators used in conjunction with the flotation process) or may be in the form of a counter-current flow (as in gas-stripping applications). Using Figure 14.7 as a process definition diagram, the

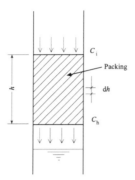

Figure 14.7
Packed column
schematic layout

gas transfer characteristics of packed columns can be derived by writing Equation (14.11) in the form

$$dc = k_L a\, dt(C_s - C)$$

Applied to a packed column, this relationship can be written as follows:

$$\frac{dc}{dh} = K_L a(C_s - C)\frac{dt}{dh}$$

$$\frac{dc}{dh} = \frac{k_L a \cdot t}{h}(C_s - C) \qquad (14.26)$$

where t is the 'hold-up' time in the column and h is the column height. It is assumed that $dt/dh = t/h$, i.e. there is no vertical variation in the liquid residence time through the column depth.

The ratio $h/(k_L a \cdot t)$ has the dimensions of length and has been found to be approximately constant for a given gas transfer process. This packed column characteristic is known as a 'transfer unit height' or *HTU*. Thus equation (14.26) can be written as

$$\frac{dc}{dh} = \frac{1}{HTU}(C_s - C) \qquad (14.27)$$

Integration yields:

$$\ln\left(\frac{(C_s - C_i)}{(C_s - C_h)}\right) = \frac{h}{HTU} \qquad (14.28)$$

where h is the packing height, C_i is the inflow dissolved gas concentration and C_h is the outflow dissolved gas concentration.

Typical HTU values (air to water transfer):

	HTU (mm)
25 mm and 38 mm spheres	285
46 mm tellerettes	405
70 mm telleretes	375
25 mm Raschig rings	500

REFERENCES

Adeney, W. E. and Becker, H. G. (1918; 1919, 1920) *Sci. Proc. R. Dublin Soc.*, **15**, 31 (1918); **15**, 44 (1919); **16**, 13 (1920).

ASCE (1984) American Society of Civil Engineers, *A Standard for the Measurement of Oxygen Transfer in Clean Water*, ASCE, New York.

Barnhart, E. L. (1969) *Proc. Am. Soc. Civ. Eng., Sanit. Eng. Div.*, **95**, SA3, 645.

Casey, T. J. (1971) *Ph.D. Thesis*, Dept Civ. Eng., University College Dublin.

Downing, A. L. (1960) *J. Inst. Public Health Engs.*, **59**, 80.

Downing, A. L., Bailey, R. W. and Boon, A. G. (1960) *J. Inst. Sewage Purif.*, **Pt. 3**, 231.

Holroyd, A (1952) *Water Sanit. Eng.*, **3**, 301.

Karmo, O. T. (1972) *M. Eng. Sc. Thesis*, Dept Civ. Eng., University College Dublin.

Lewis, W. K. and Whitman, W. G. (1924) *Ind. Eng. Chem.*, **16**, 1215.

Pasveer, A. and Sweeris, S. (1965) *J. Water Pollut. Control Fed.*, **37**, 1267.

USEPA (1989) *Design Manual: Fine Pore Aeration Systems*, Center for Environmental Research Information, Cincinnati.

15

Anaerobic Processes

15.1 INTRODUCTION

Anaerobic digestion is a biological process in which organic matter is converted to methane (CH_4) and carbon dioxoide (CO_2) by the co-ordinated activity of anaerobic microbes, mainly bacteria. The term anaerobic implies the absence of oxygen which, in practice, means the exclusion of air from the process. The process occurs widely in natural anaerobic environments such as peat bogs, marshes, lake sediments and in the rumen of cud-chewing animals. It has also long been in use for the stabilization of sewage sludge.

Methane, the major gaseous end-product of anaerobic digestion, is biochemically inert and hence its accumulation is not inhibitory to the continuation of the process. Although it was not until the third quarter of the nineteeth century that anaerobic bacteria (*Clostridia*) were dis-covered by Pasteur, anaerobic fermentations have been exploited by man for thousands of years, e.g. in wine and beer making, in animal skin preparation for tanning and retting, etc. The first direct application of anaerobic digestion to sewage solids is attributed (Metcalf and Eddy, Inc. 1991) to Louis H. Mouras of Vesoul, France, who developed a sealed cesspool (*ca*. 1860) in which it was claimed that sewage solids were liquified. The first exploitation of digester gas (biogas) is attrib-uted to Donald Cameron, who built a septic tank for the city of Exeter, England, in 1895. He collected and used the gas for lighting in the vicinity of the works. Anaerobic digestion processes were subsequently incorporated in sewage treatment systems using dual-purpose tanks combining sedimentation and digestion. The Travis hydrolytic tank (Hampton, England, 1904) and the Imhoff tank (Germany,1904) were notable examples of such systems. In the 1920s and 1930s there was considerable development in the application of anaerobic digestion for sewage sludge stabilization with the construction of many heated diges-ters in the United States and Europe, particularly for large cities.

15.2 PROCESS MICROBIOLOGY

The conversion of complex organic molecules to methane and carbon dioxide in an anaerobic digester is accomplished by a diverse community

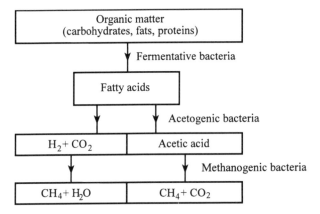

Figure 15.1
Biological
conversion of
organic matter

of interdependent microbial populations. Scientific progress towards a complete understanding of anaerobic ecology has been retarded by, amongst other things, the potential difficulties of anaerobic culture which demands an absolutely oxygen-free environment. Although there is an extensive literature on the subject, it would appear that knowledge of bacterial population dynamics in anaerobic digesters is rather limited. Bacteria are the main agents of biochemical conversions, but some anaerobic fungi and protozoa may also be involved. For many years anaerobic digestion was macroscopically described as a two-stage process. The first or acid-forming stage involved the hydrolysis of biopolymers (carbohydrates, fats and proteins) to fatty acids, while in the second stage the free fatty acids were converted to CH_4 and CO_2 by the methanogenic bacteria. In the light of present understanding of the overall metabolic process, it is more appropriately described as a three-stage process, as depicted in Figure 15.1. The first stage involves the degradation of biopolymers to fatty acids by fermentative bacteria. In the second stage the higher fatty acids are converted in part to acetic acid and in part to the H_2/CO_2 combination by the acetogenic group of bacteria. Acetic acid and H_2/CO_2 are the key substrates of the methanogenic bacteria. It is estimated that about 70% of the methane produced is derived from acetic acid. Although hydrogen is an important precursor of methane formation in digesters, measured concentrations of H_2 are usually very low, suggesting an immediate uptake by the bacteria.

The methanogens are considered to be the key process organisms and are the most environmentally sensitive group in a digester population. They are unusual in that they are composed of many species with very different cell morphologies. Digester populations can be divided into two temperature groups: the mesophilic group is active between 10°C and 40°C and the thermophilic between 40°C and 65°C.

15.3 ENVIRONMENTAL INFLUENCES ON THE DIGESTION PROCESS

The methanogens are strict anaerobes and require a reduced environment for growth—the prevailing redox potential in anaerobic cultures has been found (Hughes, 1979) to be in the range -150 to -420 mV (relative to the hydrogen electrode), being generally closer to the latter value. A dissolved oxygen level as low as 0.01 mg l^{-1} has been found (Wolfe, 1971) to completely inhibit methanogenic growth. Methane formation does not readily occur in the presence of electron acceptors such as nitrate and sulphate which, under anaerobic conditions, are reduced to nitrogen and sulphide, respectively, prior to the generation of methane.

Sulphide, which is an essential nutrient, is toxic at concentrations exceeding about 200 mg l^{-1} (Mosey, 1976). Sulphides are removed through the formation of highly insoluble metallic sulphides.

Dissolved heavy metals, notably chromium, copper, nickel, cadium and zinc, exhibit toxicity above certain critical levels. The sequestration of these metals by organic matter and their precipitation as sulphides can, however, greatly reduce the dissolved metal fraction. Table 15.1, taken from a study by Barth *et al.* (1965), shows the concentration at which the indicated heavy metals can be present in sewage without adversely affecting the anaerobic digestion of the resulting sludge. The light metal cations—sodium, calcium and magnesium—have also been found to be toxic at high concentrations (Kugelman and McCarty, 1965), with evidence of synergism (increased toxicity) and antagonism (reduced toxicity) when present in combination.

The methanogens are particularly sensitive to the pH of the digester liquor. Digestion proceeds satisfactorily in the pH range 6.6–7.6 with the optimum value close to the neutral point 7.0. Although the growth of methanogens is inhibited below pH 6.6, the fermentative bacteria continue to function until the pH drops to the region 4.5–5.0, resulting in a rapid accumulation of volatile fatty acids (VFA) in the intervening

Table 15.1 Highest metal concentration in sewage that will allow satisfactory digestion of sewage sludge

Metal	Concentration in influent sewage (mg l^{-1})	
	Primary sludge digestion	Combined sludge digestion
Chromium (hexavalent)	>50	>50*
Copper	10	10
Nickel	>40	>10*
Zinc	10	10

* Higher dose not studied.
Source: Barth *et al.* (1965)

pH range. This conjunction of low pH and VFA increase initially led to the conclusion that high concentrations of VFA were toxic to methanogens. However, many studies have shown that VFAs are not toxic to methane bacteria at the concentrations likely to occur in digesters. Van Velsen and Lettinga (1979) found no adverse affect on digestion at a VFA concentration of 5000 mg l^{-1} (as acetic acid), while Newell (1981) reported satisfactory operation of an anaerobic filter at influent VFA concentrations up to 11 800 mg l^{-1}.

Ammonia is an essential nitrogen source for anaerobic growth (C:N ratio *ca.* 30:1) but is toxic at higher concentrations. In the range 1500–3000 mg l^{-1} of ammonia–nitrogen, ammonia has been found to be inhibitory at pH above 7.4. At concentrations above 3000 mg l^{-1} ammonia–nitrogen, ammonia has been found to be toxic regardless of pH.

Other potential inhibiting agents, which may sometimes be found in waste sludges particularly, include anionic detergents, chlorinated hydrocarbons, phenols and antibiotics. Anionic detergents, which are generally of domestic origin, may cause significant inhibition (Swanwick and Shurben, 1969) if their concentration exceeds about 1.5% of the dry sludge solids. The other organic inhibitors mentioned are generally of industrial origin and should not normally be discharged into public sewers in significant amounts.

Digestion temperature has a major influence on process kinetics, as already discussed. In the mesophilic range, i.e. 10–40°C, it is generally found that maximum gas production occurs around 30–35°C, the optimum value apparently depending on feed composition and the digester population. In the thermophilic range, i.e. 40–65°C, the rate of gas production and quantity of gas produced increases with an increase in temperature up to about 60°C. Pfeffer (1979) found that thermal inhibition of methanogenic bacteria occurred at about 63°C. There is also some evidence to indicate that there may be a thermally unfavourable temperature zone between the mesophilic and thermophilic temperature ranges, i.e. between 35°C and 45°C, where erratic gas production may be encountered.

15.4 PROCESS KINETICS

The Monod model of microbial growth kinetics, as outlined in Chapter 11, can be applied to anaerobic methanogenic systems. To apply this model in design, information is required on the model parameters Y, $\hat{\mu}$, k_d and K_s, for which some values, typical for anaerobic processes, are given in Table 11.4.

In practical design situations, however, precise values for these parameters are usually not available, especially for complex wastes such as sewage sludge. For example, substrate concentration is not measured directly but in terms of non-specific parameters such as COD or BOD.

Despite this limitation, the kinetic model provides a very useful basis for process evaluation. Experimental measurements indicate approximately constant values for the coefficients Y and k_d for complex substrates such as sewage sludge. The value of Y is reported to be about 0.04 mg per mg COD removed and that of k_d to be about 0.01 d^{-1}. The value of Y reflects the low production of microbial sludge in anaerobic systems.

It is well known that anaerobic processes are very sensitive to temperature. The microbial growth coefficients $\hat{\mu}$ and K_s vary with temperature according to the following empirical correlations

$$\hat{\mu}_T = 0.08 \ e^{0.034T} \ (\text{d}^{-1}) \tag{15.1}$$

$$(K_s)_T = 90.6 \ e^{-0.106T} \ (\text{g COD l}^{-1}) \tag{15.2}$$

where T is the process temperature (°C). Using these $\hat{\mu}_T$ and $(K_s)_T$ values and a k_d value of 0.01 d^{-1}, the predicted residual COD for sewage sludge is plotted in Figure 15.2 as a function of residence time and temperature. This graphical representation of the foregoing equations underlines the key influences of temperature and residence time on process performance. It shows that the sensitivity of residual COD to residence time is high at low residence times, but becomes very low as the residence time increases. For example, inspection of Figure 15.2 shows that there is only a marginal reduction in residual COD for retention times beyond 10 days, for a digestion temperature of 35°C. The model also implies that the residual COD is independent of the influent COD concentration. The results of laboratory experiments on sewage sludge digestion (Mosey, 1976) generally confirm the above model characteristics. It was found that regardless of whether the solids content of the sludge was 3%, 6% or 9%, the proportion of volatile solids

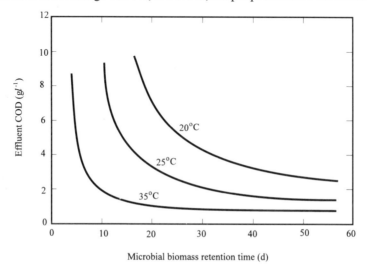

Figure 15.2
Influence of process temperature on residual COD in anaerobic digestion of sewage sludge

destroyed remained constant at about 45%, when the nominal residence times were progressively reduced from 50 d to 11 d. It seems clear therefore that the design practice of using residence times of 20+ d for completely mixed sewage sludge digesters is unduly conservative. A nominal design residence time of 12–15 d should provide an adequate safety margin to cope with variable loading.

As shown in Chapter 11, the reactor volume required to achieve a given degree of substrate conversion is, according to the foregoing kinetics, less for a plug-flow reactor than for a completely mixed reactor. Upflow anaerobic sludge blanket contact reactors, which do not have an external mixing input, rely on the kinetic energy of the influent and on biogas generation to achieve an adequate level of mixing. Flow-through systems, as used for sewage sludges, are invariably of the completely mixed type. However, it is also possible to use plug-flow reactors for sewage sludge provided there is adequate sludge recycling for seeding purposes, and an appropriate mixing system is used.

15.5 PROCESS TECHNOLOGY

Modern developments in the exploitation of methanogenic processes can be broadly divided into two categories: 'flow-through' systems, as illustrated in Figure 15.3, in which the residence time of the feed and the microbial biomass in the process unit is the same, and 'contact' systems, as illustrated in Figure 15.4, in which the feed stream is brought into contact with the active microbial biomass retained in the system.

Flow-through reactors are used for the digestion of concentrated feed materials such as animal manures or sewage sludges, which have solids concentrations typically in the range 2–10% on a dry weight basis. Sewage sludge digesters are invariably of the completely mixed type, with sludge recycling as a common design feature. Flow-through digesters of the plug-flow type have been used to a limited extent for the digestion of animal manures (Hayes *et al.*, 1979). In addition to the potential kinetic advantages already discussed, plug-flow allows the use of a simple tank configuration, thus reducing capital costs.

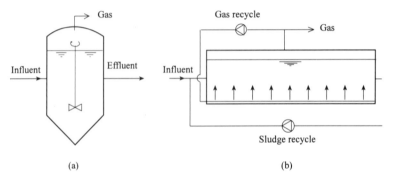

Figure 15.3
Schematic outlines of flow-through reactor configurations. (a) Completely mixed reactor; (b) plug-flow reactor

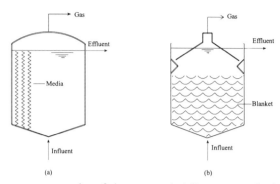

Figure 15.4
Schematic
outlines of
contact reactor
configurations. (a)
Static-media
biofilter; (b)
sludge-blanket
process

Contact systems may be of the suspended floc or attached film type. Suspended floc systems incorporate a floc-separation stage, which may be external to the system and require solids recycling, as in the aerobic activated sludge process, or may be incorporated within the system, as in the sludge-blanket reactor, shown in Figure 15.4(b). This latter system uses a process technology similar to that used in sludge-blanket clarification in potable water production.

Attached film processes operate in a flooded upflow mode, as shown in Figure 15.4(a), and may be of the static or fluidized media type. The media used in static beds may be natural stone or plastic packing, while fluidized beds incorporate a fine-grained inert medium such as sand. Fluidized bed reactors require a high level of recirculation to achieve a fluidization upflow velocity.

15.6 PROCESS DESIGN

The overall technical objective of process design may be said to be the provision of an optimal growth environment for the microbial population of the digester. The parameters over which the designer can exercise control include tank volume and shape, operating temperature, mixing input and, in some instances, the influent solids concentration.

From the foregoing discussion of process kinetics, it will be clear that the reactor design volume is a function of the influent composition, the operating temperature, the microbial concentration in the digester and the required effluent quality. The active solids residence time θ_s has already been identified in the discussion of kinetics as a key process parameter. In flow-through systems, hydraulic and solids residence times are the same. A design residence time of 12–15 d has been found to be adequate for the mesophilic digestion (30–35°C) of complex substrates such as sewage sludge and animal manures in flow-through digesters. The required tank volume is simply the product of inflow rate and residence time. The active solids residence time is an equally valid design parameter for anaerobic contact systems. However, as discussed in Chapter 12, it is more usual to use the sludge loading L_s or the specific substrate removal rate L_s as the design parameter for both aerobic and anaerobic contact systems:

$$L'_s = \frac{1}{\theta_s(X_o + Y\Delta S(1 + (1 - f_b)(k_d\theta_s))/(1 + k_d\theta_s))} \text{ kg}$$

COD removed per kg MLVSS.d   (12.2)

For an anaerobic contact process operating at 35°C, the following set of parameter values may be inserted into equation (12.2):

$$\begin{array}{ccc} X_o = 0.1 & Y = 0.04 & \Delta S = 0.8 \\ f_b = 0.8 & k_d = 0.01 \text{ d}^{-1} & \theta_s = 10 \text{ d} \end{array}$$

giving

$$L'_s = 1.2 \text{ kg COD removed per kg MLVSS.d}$$

An important difference between aerobic activated sludge processes and the corresponding anaerobic suspended floc contact systems is that the latter can be operated at much higher MLSS concentrations—in the range 40–60 kg m^{-3} for sludge-blanket-type anaerobic reactors (Lettinga *et al.*, 1979). When this MLSS value is combined with the above estimated value for L_s, it is clear that very high space-loading rates can be applied to such reactors. Sludge-blanket-type reactors have been sucessfully operated at organic space-loading rates up to 40 kg COD d^{-1} m^{-3}, at a temperature of 30°C, achieving a COD reduction in the range 70–90%. Newell (1981) reported an 80% COD removal from the liquid fraction of pre-fermented pig manure, digested at 28°C in an upflow static filter, loaded at about 19 kg COD d^{-1} m^{-3}. Thus, a key goal in the design and operation of anaerobic contact processes is the development of as large a microbial population density as possible in the digester. The microbial population density in an attached film reactor is related to the available surface area for attachment, which, for rounded granular media, increases as the reciprocal of the grain size. Thus, the use of fine sand instead of a coarse granular medium in an anaerobic filter provides the capacity for the accumulation of a greatly increased microbial population density and hence a greater treatment capacity per unit volume. The operation of such a system in an expanded or fluidized upflow mode prevents clogging, which is a potential problem in static filters. On the basis of laboratory work, Jewell (1979) reported that expanded bed digesters can be used to treat wastes ranging from dilute settled domestic sewage to cow manure at 2% solids. A disadvantage of fluidized systems is the high flow rate required for fluidization.

The temperature of operation is a basic design parameter, influencing process kinetics and the overall degree of gasification, as shown in Figure 15.2. It also directly affects the energy balance of the process, which is a critical factor where the digester gas is used for heating or where net energy production is the prime objective in the process design. The required heat input is made up of two components: that required to raise the influent temperature to the operating level for the digester, and the make-up heat required to offset the heat loss from the digester vessel:

$$\text{heat input } H_i = \rho \, Q S_h \, \Delta T_1 + A U_a \, \Delta T_2 \qquad (15.3)$$

where ρ and S_h are the density and specific heat capacity, respectively, of the digester influent, ΔT_1 is the temperature difference between the influent and the digester contents, ΔT_2 is the temperature difference between the digester contents and the surrounding air, A is the total surface area of the digestion vessel and U_a is its average overall heat transfer coefficient. In mesophilic applications, ΔT_1 typically has a design range of 10–35°C, while the corresponding range for ΔT_2, under temperate climatic conditions, is about 5–40°C. It will be obvious from equation (15.3) that the heat input requirements can be reduced by reductions in the influent volume, the digester surface area and the thermal conductivity of the construction. Thus, in a particular application, an increase in influent solids from 2% to 4%, which reduces the volume by a factor of about two, effectively halves the heat input requirement. The importance of thermal insulation is demonstrated by an illustrative example in Figure 15.5. The plotted graphs relate to a free-standing concrete cylindrical tank with a flat floor and roof, having a constructional thickness of 200 mm throughout. At a typical flow-through digester design retention time of 15 d, the use of 50 mm thick glass wool insulation on the walls and roof reduces the heat required to about half that required for the insulated system. The optimum degree of insulation can be determined from economic evaluation, taking into account the respective costs of heat energy and insulation.

Mixing (Verhoff *et al.*, 1974) is an area of digester design that requires special consideration. It is essential to ensure that local dead spaces are not allowed to develop in a digester, reducing the available active volume. Equally, it is desirable to have a uniform temperature throughout the digester. A gentle level of mixing is adequate to achieve these

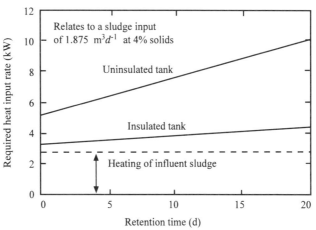

Figure 15.5
Influences of insulation and retention time on digester heating requirements.

Digester/air temperature difference 30°C
Digester/feed sludge temperature difference 30°C

goals. In contact systems the required level of mixing is provided by the upflow velocity of the influent and is also aided by the gas generated within the system. In flow-through systems, an external energy input is usually required, which may be in the form of mechanical agitation, sludge recirculation or gas recirculation. The latter is considered (Konstandt, 1976) the most appropriate method for digesters, being complementary to the mixing effect of the gas generated in the sludge bulk. It offers the designer flexibility in respect of tank geometry, being readily adaptable to any shape of tank. The mixing effectiveness of shaft-mounted paddles, turbines or propellers, on the other hand, is dependent on the flow pattern generated by these devices, which is influenced to a large extent by tank geometry and by the viscous nature of the anaerobic mixed liquor. There appears to be no established design norm in respect of power input per unit volume for digester mixing at the present time. Indeed, it may well be that at high solids concentrations there may be adequate self-mixing through gas evolution in the system. In summary, appropriate mixing to ensure full digester volume utilization is important, but intense mixing is not advantageous.

As already pointed out, in the case of substrates such as sewage sludges and animal manures, the influent suspended solids is an important process variable, over which the designer may have control. The volume of a digester, for a given residence time and influent solids mass, bears an inverse relation to the solids concentration of the influent. Experience has shown that complex organic mixtures such as sewage sludges and animal manures can be digested, without significant loss in efficiency, at solids concentrations up to 9% or 10%. Pre-thickening to remove of excess water is therefore likely to be an economically attractive design option for many slurry-type influents, particularly sewage sludges. Not only is the required digester volume thereby reduced, but so is the heat input requirement. Checks should be made, however, to ensure that pre-thickening of the digester feed does not concentrate inhibitory agents to a toxic level.

15.7 DIGESTER GAS

Digester gas consists mainly of methane and carbon dioxide, in proportions that vary with the composition of the substrate being degraded and in output quantities proportional to the amount of organic matter mineralized. Small amounts of other gases, notably nitrogen, hydrogen and hydrogen sulphide, may also be present. Average values for digester gas yield and composition are given in Table 15.2.

It will be noted that the yield of methane from fats is about twice that from carbohydrates. Methane is the valuable component of digester gas. It is therefore useful to express gas yield in methane terms as

m^3CH_4 per kg organic matter converted or as m^3CH_4 per kg COD removed. Some reported methane yield values are given in Table 15.3. The methane yield from liquid wastes, treated in contact processes, has been reported (Hayes *et al.*, 1979) to vary in the range 0.3–0.5 m^3 per kg COD removed.

Table 15.2 Biogas composition and yield

Substrate	Gas yield and composition		
Carbohydrates	0.8 m^3 kg^{-1}	50% CH_4	50% CO_2
Proteins	0.7 m^3 kg^{-1}	70% CH_4	30% CO_2
Fats	1.2 m^3 kg^{-1}	67% CH_4	33% CO_2

Source: Konstandt (1976)

Table 15.3 Typical digestion yield at 35°C

Substrate	Methane yield (m^3 per kg organic matter removed)
Mixed sewage sludge	0.8–0.9
Dairy manure	0.5–0.6
Pig manure	0.4–0.5

It will be clear from the discussion on kinetics that the extent of conversion of organic matter to gaseous end-products depends on the operating temperature and the residence time in the digester. At 35°C some 40–50% of the volatile fraction of sewage sludges is removed in retention times of 12–15 d. The yield of methane from sewage sludge under these conditions, on a per capita basis, has been reported (Metcalf and Eddy, Inc. 1991) to be 12–16 l per capita per day for primary sludge and about 20 l per capita per day for combined primary and secondary sludges. Pfeffer (1979) found that the anaerobically biodegradable fraction of solid refuse increased from 30% of the volatile solids at 35°C to 55% at 60°C.

Digester gas is usually stored in low-pressure, variable-volume gasholders at a pressure in the range 100–300 mm water gauge. The required storage volume depends on the usage pattern. Utilization for purposes other than process heating is generally determined by economic considerations. Potential uses include heating of local buildings, use as an engine fuel to drive electrical generators or vehicles, local industrial use and discharge to a municipal gas network after appropriate upgrading.

It scarcely needs stressing that detailed attention to the safety aspects of gas handling is essential both at the design stage and during operation.

15.8 POTENTIAL USES

The extent to which anaerobic processes will displace aerobic processes for organic waste stabilization in the future will depend to a large extent on enonomic factors, which are likely to be influenced to an increasing degree by energy costs. The organic waste residues in question include domestic sewage and sewage sludge, solid refuse, organic industrial wastewaters, animal manures and crop residues. Anaerobic digestion processes are likely to be applied to an increasing extent in the case of three of these residues, namely sewage sludge, concentrated organic industrial wastewaters and animal manures.

Mesophilic digestion processes having a residence time of 15–25 d, have long been used for sewage sludge stabilization, particularly in treatment plants of medium to large size (>30 000 PE). Aerobic stabilization is generally preferred for smaller plants.

Factors such as the cost of energy and the development of more efficient and less costly digester configurations will probably lead to the increased use of anaerobic digestion in the sludge disposal field. Modifications of current design practice, which would lead to significant reductions in process capital cost, include:

- reduction in design sludge residence time from 20–25 d to 12–15 d;

- use of plug-flow digesters of simple-shape, mixed by recycled gas (such designs have yet to be developed);

- reduction of sludge volume by thickening prior to digestion.

In the field of industrial wastewater disposal, anaerobic contact processes would appear to have great potential for partial treatment of warm concentrated biodegradable wastes. The particularly attractive features of the process in this field of application are:

- very little surplus sludge is produced;

- no external heat input is required;

- usable energy in the form of methane gas is produced.

Summing up, therefore, it seems probable that the combined influences of improved technology and energy cost inflation could lead to a growing use of anaerobic digestion processes. However, the inherent operational sensitivity of the process will remain a serious constraint on this expansion. This applies particularly to situations where there is inadequate control over influent quality, as is sometimes the case with sewage sludges, particularly at small municipal wastewater treatment works receiving significant industrial discharges. In such situations, the environmentally more robust aerobic processes will continue to be preferred.

REFERENCES

Barth, E. F., Moore, W. A. and McDermott, G. N (1965) *Interaction of Heavy Metals on Biological Sewage Treatment Processes*, US Dept of Health, Education and Welfare.

Jewell, W. J. (1979) *1st Internat. Symp. on Anaerobic Digestion*, University College, Cardiff.

Hayes, T. D., Jewell, W. J., Dell'Orto, S., Fanfoni, K. J., Leuschner, A. and Sherman, D. F. (1979) *1st Internat. Symp. on Anaerobic Digestion*, University College, Cardiff.

Hughes, D. E. (1979) *1st Internat. Symp. on Anaerobic Digestion*, University College, Cardiff.

Konstandt, H. G. (1976) *Microbial Energy Conversion*, ed. H. G. Schlegel and J. Barnea, Erich Goltze KG, Gottingen.

Kugelman, I. J. and McCarty, P. L. (1965) *J. Water Pollut. Control Fed.*, **37**, 97.

Lettinga, G., van Velsen, A. F. M., de Zeeuw, W. and Hobma, S. W. (1979) *1st Internat. Symp. on Anaerobic Digestion*, University College, Cardiff.

Metcalf and Eddy, Inc. (1991); *Wastewater Engineering*, 3rd edn, McGraw-Hill Book Co., New York.

Mosey, F. E. (1976) *Water Pollut. Control*, **75**, 10.

Newell, P. J. (1981) *Energy Conservation and the Use of Solar and other Renewable Energies*, ed. F. Vogt, Pergamon Press, Oxford.

Pfeffer, J. T. (1979) *1st Internat. Symp. on Anaerobic Digestion*, University College, Cardiff.

Swanwick, J. D. and Shurben, D. G. (1969) *Water Pollut. Control*, **68**, 190.

van Velsen, A. F. M. and Lettinga, G. (1979) *1st Internat. Symp. on Anaerobic Digestion*, University College, Cardiff.

Verhoff, F. H., Tenney, M. W. and Echelberger, W. F. (1974) *Biotech. Bioeng.*, **16**, 757.

Wolfe, R. S. (1971) *Adv. Microb. Physiol.*, **6**, 107.

Related reading

Bruce, A. M., Kouzeli-Katsiri, A. and Newman, P. J., eds (1986) *Anaerobic Digestion of Sewage Sludge and Organic Agricultural Wastes*, Elsevier Applied Science Publishers, London.

Demuynck, M., Nyns, E. J. and Palz, W. (1984) *Biogas Plants in Europe, A Practical Handbook*, D. Reidel Publishing Company, Dordrecht.

Ferrero, G. L., Ferranti, M. P. and Naveau, H., eds (1984) *Anaerobic Digestion and Carbohydrate Hydrolysis of Waste*, Elsevier Applied Science Publishers, London.

Ferranti, M. P., Ferrero, G. L. and L'Hermite, P., eds (1987) *Anaerobic Digestion: Results of Research and Demonstration Projects*, Elsevier Applied Science Publishers, London.

16

Sludge Processing and Disposal

16.1   INTRODUCTION

Sludge is the relatively concentrated suspension into which the solids fraction of a water is concentrated in the course of purification. Sludges are derived from the processes of chemical coagulation and softening at waterworks and from the preliminary, primary, secondary and tertiary stages of wastewater treatment. Most of these sludges are of an unstable organic nature and readily undergo active microbial decomposition with consequent generation of nuisance odours. They all have the common characteristic of a high water content, usually greater than 95% by weight. Sludges derived from wastewaters containing domestic sewage or animal excreta may contain significant concentrations of pathogenic organisms. In this chapter, sludge composition and quantities generated are considered, as well as the available options for processing and disposal.

16.2   WATERWORKS SLUDGE

Waterworks sludge is derived almost entirely from the chemical coagulation of surface waters to effect the removal of colour and turbidity, using either aluminium or iron salts as chemical coagulants. Significant quantities of calcium carbonate sludge are produced in water softening by the lime/soda precipitation process. This process, however, is not much used in potable water treatment systems at present.

16.2.1   Characteristics of alum sludge

Alum sludge is a bulky gelatinous suspension whose solids fraction consists of aluminium hydroxide floc, fine inorganic particles, colour colloids and other organic debris. The sludge, as discharged from clarifiers, usually has a solids content in the range 0.3–1.5% dry matter by weight. It normally has a relatively low biodegradable organic fraction, typically having a 5-day biochemical oxygen demand (BOD_5) in the range of 30–100 mg l^{-1} (Albrecht, 1972). The residual solids, following sedimentation, are removed by sand filtration and are discharged from filters by back-washing at a high flow rate. The back-wash stream usually has an average solids concentration in the range 0.004–0.1% dry matter by weight while its BOD_5 is usually less than 5 mg l^{-1}.

Waterworks alum sludges are predominantly inorganic and relatively stable. However, the given BOD_5 range indicates that they have sufficient biodegradable material to cause anaerobic conditions within the sludge bulk.

16.3 SEWAGE SLUDGE

Sewage sludge is the relatively concentrated suspended solids residue resulting from the purification of municipal and industrial wastewaters. It includes primary sludge, which is composed of the settleable solids fraction of raw wastewater, as separated by sedimentation or flotation processes, and secondary sludges, which are derived from biological or physico-chemical treatment processes.

The solids composition of raw municipal wastewaters is quite variable, being influenced by the type of sewer collection system (whether combined or separate) and the relative contributions from domestic and industrial sources. The approximate division of municipal wastewater solids into organic and inorganic fractions, and their further division into settleable, non-settleable and dissolved components, is shown in Table 16.1.

The preliminary phase of sewage treatment normally consists of screening and grit separation. Screening removes so-called 'gross' solids from the flow. Sometimes screenings are macerated and returned to the flow. Alternatively they may be disposed to tip or buried on site. Grit is the heavy inorganic particulate fraction of sewage, which separates readily from the lighter organic settleable solids. Removal at a preliminary stage is essential to avoid its accumulation in downstream process

Table 16.1 Average composition of sewage solids expressed as percent by weight

Solids category	Suspended settleable	Suspended non-settleable	Dissolved	Total
Inorganic	10.6	5.3	21.1	37.0
Organic				
Carbohydrates	6.8	3.3	10.5	20.6
Fats	7.4	3.6	8.0	19.0
Proteins	6.8	3.6	13.0	23.4
Organic total	21.0	10.5	31.5	63.0
Total	31.6	15.8	52.6	100.0

Source: Popel (1963).

units. The grit load may typically vary in the range 25–250 mg l^{-1} (White, 1970), being particularly high in combined sewage at times of heavy rainfall. When grit is washed free of organic matter, it becomes an inert granular residue which can be used as a construction material or disposed to landfill.

Primary treatment of sewage normally involves simple sedimentation to remove settleable solids. Secondary treatment consists of a biological process—biofiltration or activated sludge—followed by sedimentation. Tertiary treatment may consist of sand filtration to remove residual suspended solids and/or chemical precipitation of phosphorus. Typical sludge yields from sewage treatment processes are given in Table 16.2. These ranges should be taken as a rough guide only, as the production of sludge varies considerably from works to works. The lower limits of the ranges may be taken as typical for domestic sewage with the upper limits indicative of a significant industrial contribution.

Table 16.2 Sludge production in sewage treatment

Process	Quantity ($g\,cap^{-1}\,d^{-1}$)	Solids content (% by weight)	Volume ($l\,cap^{-1}\,d^{-1}$)
Primary sedimentation	45–55	5–8	0.6–1.1
Biofiltration of settled sewage	13–20	5–7	0.2–0.4
Standard rate activated sludge pre-settled sewage	20–35	0.75–1.5	1.3–4.7
Extended aeration raw sewage	22–50	0.75–1.5	1.7–6.7
Tertiary sand filtration	3–5	0.01–0.02	15.0–50.0
Phosphorus precipitation (Al or Fe)	8–12	1–2	0.4–1.2

Source: Popel (1963); Gale (1971).

16.3.1 Sewage sludge characteristics

Primary sludge is a grey slimy suspension with an offensive odour, typical of active putrefaction. Since it contains wastes of enteric origin it may contain significant numbers of pathogenic organisms. Activated sludge is a flocculent suspension with a characteristic inoffensive earthy odour, when fresh. It is essentially a microbial biomass consisting mainly of bacteria but also including fungi and protozoa. The stability of activated sludge is a function of the loading rate of the process in which it has been produced. Extended aeration activated sludges are stable while standard rate and high rate activated sludges readily become septic in the absence of aeration. Sewage-derived activated sludge may contain some surviving pathogenic organisms (van Gils, 1964). In general, biofiltration sludges are similar to activated sludges.

Sewage sludges exhibit non-Newtonian flow and have thixotropic properties. Their specific gravities reflect the specific gravities of their solids fractions and also depend on solids concentration.

16.4   INDUSTRIAL WASTEWATER SLUDGES

The principal industrial activities which give rise to sludge- producing wastewaters are:

- Food and related process industries, including milk, meat, fish, fruit and vegetables processing.

- Beverage industries including malting, brewing, distilling and soft drinks production.

- Pulp and paper, chemicals, biochemicals, textiles production.

- Mining and quarrying.

With the exception of mining and quarrying, where the derived sludge residues are entirely inorganic, and certain relatively toxic residues from the the chemical/biochemical sectors, industrial wastewaters are mainly of a biodegradable nature and are amenable to the same range of treatment processes as municipal sewage. The distribution of solids by size and composition varies considerably from industry to industry. The general physical characteristics of the primary and secondary sludges derived from processing organic industrial wastewaters are broadly the same as those described for sewage sludges.

16.5   PROCESSING OF SEWAGE SLUDGE

Water and wastewaters sludges, as discharged from the process units in which they are produced, vary greatly in composition and water content. The degree to which they are subsequently processed depends largely on their disposal destination, which may be to agriculture, landfill or incineration. The range of process options is outlined in Figure 16.1. The optimal process route for a particular sludge is determined by technical, economic and environmental considerations.

For discussion purposes, sludge treatment processes can be conveniently divided into two groups: (a) solids/water separation processes and (b) sludge solids stabilization processes.

16.6   SOLIDS/WATER SEPARATION PROCESSES

The water fraction of sludge can be divided into three categories: (1) free water, which is not intimately associated with sludge solids; (2) capillary

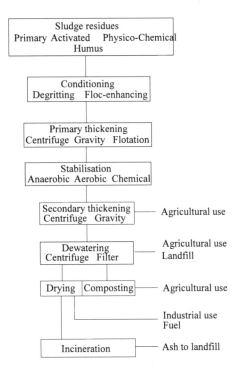

Figure 16.1
Sludge processing
flow sheet

and boundary layer water retained by surface forces; and (3) intracellular and chemically bound water.

The ease with which the free water can be drained from a sludge depends on its hydraulic permeability. Sludges that have a flocculent solids fraction have relatively large drainage channels through which the water can escape while the opposite holds for sludges of a gelatinous nature. In general, organic sludges have a high affinity for water and require artificial flocculation ('conditioning') prior to thickening and dewatering. Flocculation causes an aggregation of solids particles, in consequence of which a considerable fraction of the water is transformed into the free drainage category.

Capillary and surface-held water can only be removed by the application of pressure gradients which exceed the counter-gradients generated by holding forces such as surface tension.

The following laboratory measurements of solids/water separability are used: sludge volume index (*SVI*), specific resistence to filtration (*SRF*) and capillary suction time (*CST*).

SVI is the sludge volume, as ml, per 1 g of sludge solids after 1 l of sludge is allowed to settle for 30 minutes in a standard graduated cylinder. It essentially measures sludge settleability and is particularly relevant to sedimentation and thickening processes. An activated sludge with good settling properties typically has an *SVI* less than 100 ml g^{-1}, while a 'bulking' sludge may have an *SVI* value in excess of 200 ml g^{-1}.

Sludge *SRF* is a measure of the impermeability of a sludge layer deposited on a filter medium to which a vacuum has been applied. The rate of filtration can be expressed as follows:

$$\frac{dV}{dt} = \frac{p}{\mu} \cdot \frac{A}{R} \qquad (16.1)$$

where dV/dt is the filtration rate, p is the total pressure difference across the cake and the filter medium, μ is the viscosity of the liquid, and A is the filter plan area.

This equation defines R, the total resistance to flow of the cake and the medium. The total flow resistance may be divided into the cake resistance R_c and the medium resistance R_m, and equation (16.1) may be written as follows:

$$\frac{dV}{dt} = \frac{p}{\mu} \left(\frac{A}{R_c + R_m} \right) \qquad (16.2)$$

R_c is not a constant because, as filtration proceeds, the depth of the cake increases as does its hydraulic resistance. It has been found convenient to directly correlate this increase in resistance to w, the mass of solids deposited per unit cake area. Hence equation (16.2) becomes:

$$\frac{dV}{dt} = \frac{p}{\mu} \frac{A}{(rw + R_m)} \qquad (16.3)$$

This equation defines the *SRF* value r as the hydraulic resistance of a cake having unit mass of dry solids per unit area of filtration surface, with units of m kg^{-1}.

To use equation (16.3) for practical purposes, w is expressed in terms of the volume of filtrate per unit area by writing $w = cV/A$, where c is the suspended solids concentration in the sludge sample being filtered. Substitution in equation (16.3) yields:

$$\frac{dV}{dt} = \frac{p}{\mu} \frac{A^2}{(rcV + R_m A)} \qquad (16.4)$$

If the assumption is made that the terms in equation (16.4), other than V and t, are constants, the equation can be integrated to give

$$\frac{t}{V} = \left(\frac{\mu r C}{2A^2 p} \right) V + \left(\frac{\mu R_m}{Ap} \right) \qquad (16.5)$$

The data required for computing *SRF*, using equation (16.5), are obtained from a sludge filtration test (Vesilind, 1975). The ratio t/V is plotted as a function of the filtrate volume V to give a straight-line plot of slope $(\mu rc/2A^2 p)$, in accordance with equation (16.5). The *SRF* value r is computed from the measured slope of the plotted line.

Typical *SRF* values for a variety of sludges are given in Table 16.3.

CST is the time taken in seconds for the wetting front generated by a sludge sample in contact with thick blotting paper to travel 10 mm along the paper. It is measured in a standard apparatus (Baskerville and Gale, 1968), permitting rapid evaluation. The *CST* is a measure of sludge impermeability in a low hydraulic gradient environment. Because of its simplicity and rapidity, it is a very useful test for evaluating the relative merits of the conditioning chemicals used in sludge thickening and dewatering.

Table 16.3 Specific resistance of sewage sludges and other materials

Suspension	Specific resistance to filtration (10^{13} m kg^{-1})	Filtration pressure (k N m^{-1})
Raw sewage sludges (mixed primary and secondary) from 8 works	10–20	49
The same sludges after anaerobic digestion	3–30	49
Activated sewage sludges	0.1–1000	49
Thixotropic mud	15	550
Gelatinous iron $(OH)_3$	1.5	173
Gelatinous aluminium $(OH)_3$	2.2	173
Gelatinous magnesium $(OH)_2$	0.3	173
Colloidal clay	0.5	173
Ferric oxide (pigment)	0.08	173
Precipitated $CaCO_3$	0.02	173
Ordinary kieselguhr	0.012	173
Free filtering kieselguhr	0.00017	
Shale from coal washing	0.2	49
Sludges from		
stream peeling of carrots	4.6	49
Lime neutralization of mine water	0.30	49
Vegetable tanning	0.15	49
Alum treatment of water	5.3	49

Source: Gale (1971).

16.6.1 Sludge conditioning

The water characteristics of sludges can be improved by chemical and thermal conditioning processes as well as by elutriation. Chemical conditioners include the salts of trivalent iron and aluminium as well as organic polyelectrolytes. They effect a breakdown of the natural barriers to particle aggregation or flocculation. Cationic polymers have been found (Baskerville *et al.*, 1978) to be particularly effective in this regard, using doses ranging from 2 to 5 kg per tonne of dry solids. The optimal chemical dose must be determined empirically for a particular sludge. Thermal conditioning of sludge may be by heat treatment or freezing (Vesilind, 1975). In heat treatment, the sludge is brought to about 200°C for 20–30 min at a pressure of about 18 atm. In addition to reducing *SRF*, this also effects sterilization. One disadvantage is that some organic matter is resolubilized under the extreme conditions. A freeze/thaw cycle also greatly improves sludge drainability. Thermal conditioning processes have a high energy demand and are therefore unlikely to enjoy widespread use in the future.

Elutriation is a term applied to the washing of sludge to remove fines and to lower its alkalinity, thereby improving its dewatering. It is sometimes applied to anaerobically digested sludges prior to dewatering. The recycling of sludge fines in the elutriate has been found (Vesilind, 1975) to cause problems in upstream processes.

Sludges may be conditioned at the thickening and/or dewatering stages.

16.6.2 Sludge thickening

Thickening Theory

Sludge thickening is a solids concentrating process in which the readily separated water is removed either by sedimentation or flotation. Thickening produces a slurry or concentrated sludge and is conventionally distinguished from 'dewatering' which produces a sludge 'cake', i.e. a residue with dry solid handling characteristics. Thickening can effect a considerable volume reduction; for instance, an increase in solids concentration from 1% to 4% by a thickening process reduces the sludge volume to a quarter of its original volume.

Gravitational thickening is the separation of water from sludge by particle sedimentation. It is particularly applicable to dilute sludges, where very large volume reductions can be achieved, as illustrated in Figure 16.2. A schematic outline of a continous thickener is shown in Figure 16.3. The design of such a unit is based on the concept of solids flux or downward movement of solids within the thickening unit.

According to the Kynch theory (Chapter 2), which is based on the assumption that the settling rate is a function of concentration only, the settling rate at any concentration can be obtained from a single test. In practice, however, it has been found that the Kynch theory does not hold for highly compressible materials such as wastewater sludges. The

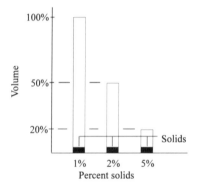

Figure 16.2
Volume reduction
by thickening

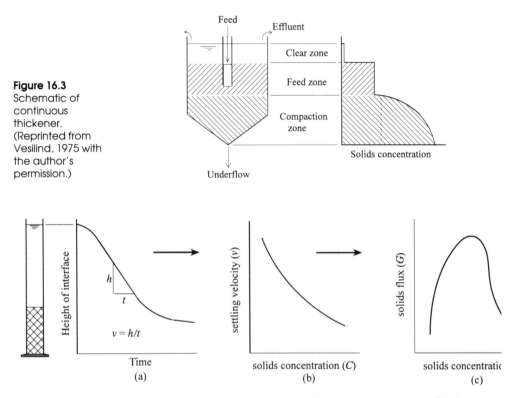

Figure 16.3
Schematic of
continuous
thickener.
(Reprinted from
Vesilind, 1975 with
the author's
permission.)

Figure 16.4 Development of the relation between solids concentration and solids flux

settling rate at a particular concentration must therefore be determined
by carrying out a settling test at that concentration, yielding a settling
curve similar to that shown in Figure 16.4(a). The slope of the linear
section of the curve gives the settling rate for that concentration. A
typical settling rate/concentration correlation is shown in Figure
16.4(b). The product of concentration and settling rate gives the solids
flux, expressed as kg m^{-2} h^{-1}. A typical solids flux/concentration
correlation is shown in Figure 16.4(c).

In a continuous thickener the total solids flux is the sum of two
components, one due to the settling rate of the sludge and the other
due to its bulk downward movement. Thus, in a continuous thickener
the total flux G_i at any level i is the sum of two components:

$$G_i = C_i(v_i + u) \tag{16.6}$$

where C_i is the concentration at level i, v_i is the settling velocity at
concentration C_i and u is the bulk downward velocity. A typical total
flux curve is shown in Figure 16.5.

Under steady operation of a continuous thickener of plan area A, the
underflow rate can be equated to the total flux at any level i:

$$G_i A = Q_u C_u = A u C_u \tag{16.7}$$

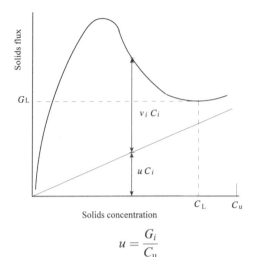

Figure 16.5
Typical total flux
curve

$$u = \frac{G_i}{C_u} \qquad (16.8)$$

where Q_u is the underflow rate. By combining equations (16.6) and (16.8) the following is obtained:

$$G_i = \frac{C_i v_i}{\left(1 - \dfrac{C_i}{C_u}\right)} \qquad (16.9)$$

For a particular value of underflow concentration C_u there is a limiting flux G_L at sludge concentration C_L, that determines the minimum thickener area required. This critical flux rate can be determined graphically from the batch flux curve, as shown in Figure 16.6.

It follows from the geometry of Figure 16.6 that:

$$\frac{G_L}{v_i C_i} = \frac{C_u}{C_u - C_i} \qquad (16.10)$$

from which it can be shown that

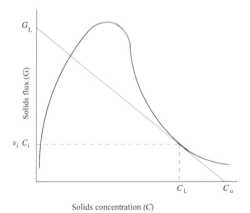

Figure 16.6
Graphical
determination of
critical flux G_L.
Source: Vesilind
(1974)
Reproduced by
permission of Ann
Arbor Science

$$G_{\mathrm{L}} = \frac{C_i v_i}{\left(1 - \dfrac{C_i}{C_{\mathrm{u}}}\right)}$$

which is identical to equation (16.9). Also

$$A = \frac{QC}{G_{\mathrm{L}}} \qquad (16.11)$$

where Q is the sludge output rate ($\mathrm{m^2\ d^{-1}}$) and C is its concentration ($\mathrm{kg\ m^{-3}}$).

Thickening Technology

Two methods of thickening are commonly employed: gravity thickening and flotation thickening. A schematic outline of a continuous gravity thickener is shown in Figure 16.3. Gravity thickener design is based (Dick, 1972; Fitch, 1975) on the concept of solids flux, expressed as $\mathrm{kg\ h^{-1}\ m^{-2}}$ of thickener plan area. Typical design values for various sludges are given in Table 16.4. The depth of the unit must accommo-

Table 16.4 Typical solids flux for gravity thickeners

Type of sludge	Solids flux range ($\mathrm{kg\ h^{-1}\ m^{-2}}$)
Activated sludge	0.6–1.0
Trickling filter humus	1.4–1.8
Raw primary sludge	4.0–5.0
Raw primary and activated sludge	1.5–4.0
Pure-oxygen activated sludge	1.0–2.0

date a clearwater zone, a feed zone and a compaction zone, as shown in Figure 16.3. Experimental evidence indicates that the effect on ultimate concentration of increasing the compaction zone thickness beyond 0.9 m is insignificant. An overall depth of 2.5–3.0 m should therefore prove satisfactory. Wherever possible, the thickening characteristics of a sludge should be assessed experimentally prior to design. Slowly rotating paddles or picket fences are sometimes used in gravity thickeners as an aid to water release from the compaction zone. In some instances (Lockyear, 1977) they have been found to effect a considerable increase in the solids concentration of thickened sludge. Typical solids concentration in gravity-thickened sludges are set out in Table 16.5.

Light flocculent sludges, such as activated sludge and waterworks alum sludge, can be thickened by dissolved air flotation processes (Sarfert, 1976; Bratby and Marais, 1977). A schematic outline of the flotation process is shown in Figure 5.4, Chapter 5. Thickening is accomplished in the lighter-than- water sludge float owing to the downward movement of water in accordance with the pressure gradient in the air/particle/water mixture. Since the concentration of solids varies within the depth of the float—being highest at the upper surface—it is normal to operate the scraping mechanism on an intermittent basis, removing the top layer with each pass.

Table 16.5 Solids concentration of gravity-thickened sludges

Type	Solids concentration after gravity thickening (%)
Mixed primary and industrial sludge (coal)	10–30
Primary sludge, volatile matter 65%	5–12
Primary + activated sludge, SVI 100 ml g^{-1}	4–6
Primary + activated sludge, SVI 100 ml g^{-1}	6–11
Sludge of extended aeration tank	3–5
Primary and trickling filter sludge	7–11
Primary digested sludge	8–14
Activated digested sludge	6–9
Heat-treated activated	10–15

Source: Kalbskopf (1972).

The basic design parameters for flotation thickeners are:

(1) solids flux rate (kg m^{-2} h^{-1}), on the basis of which, the tank surface area is determined, and

(2) the air/solids ratio, which determines the required air release rate.

The tank depth is usually in the range 2.0–2.5 m. Air/solids ratios are typically in the range 0.02–0.05 w/w, while the solids flux rate may be of the order of 10 kg m^{-2} h^{-1}. The design values for a particular application are best determined experimentally at laboratory or pilot plant level.

16.6.3   Sludge dewatering

Dewatering in this context means the removal of water to such a degree that the remaining sludge residue effectively behaves as a solid for handling purposes. The minimum solids content at which this is achieved can vary between 16% and 30%. The wide spread of this value range is due in part to the fact that sludges that have a low SRF tend to lose their fluidity at lower solids levels than those with high SRF values.

Dewatering can be accomplished by spreading on drying beds, vacuum filtration, pressure filtration and centrifugation.

Open air sand beds may be used for the dewatering of stabilized sludges at small treatment works. They usually consist of a 150–200 mm layer of fine sand, supported on a graded sand/gravel underdrainage system, contained within a shallow concrete tank structure. Sludge is discharged onto the sand to a depth of 150–300 mm and is left until it

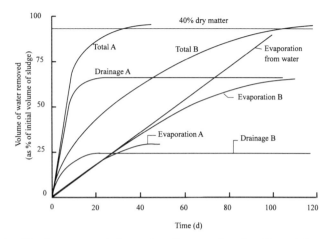

Figure 16.7
Effect of sludge SRF on rate of dewatering on drying beds. Sludge A, $SRF = 2 \times 10^{13}$ m kg^{-1}; Sludge B, $SRF = 30 \times 10^{13}$ m kg^{-1}. Source: Swanwick and Baskerville (1966)

can be lifted as a semi-dry cake. Dewatering is effected by passive drainage and evaporation, the relative amounts removed by each process depending on the sludge permeability, as shown in Figure 16.7. Since most drying beds are uncovered, the effects of rainfall must be taken into account. Experimental investigations (Swanwick, 1972) indicate that some 50–60% of the rain falling on sludge is absorbed and must later be evaporated, the remainder draining away. The solids loading to be adopted in design depends therefore on the sludge drainability and the prevailing climate. In the northern United States, recommended solids loadings are 50–120 kg m^{-2} yr^{-1} for open beds and 60–200 kg m^{-2} yr^{-1} for covered beds (WPCF, 1959).

In the United Kingdom, values of 0.33–0.50 m^2 cap^{-1} have been recommended (Swanwick, 1972), corresponding to an estimated 30–45 kg m^{-2} yr^{-1}.

Dewatering of sludge by vacuum filtration has been widely practised. A schematic outline of the process is shown in Figure 16.8. The main feature is a slowly rotating drum covered by a filter cloth. As it rotates, partly immersed in the sludge, an applied suction draws water inwards leaving a sludge layer attached to the drum surface. This layer continues to be dewatered until it reaches the discharge zone of the cycle where the

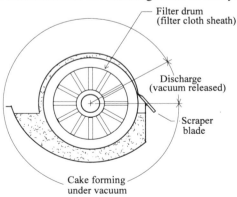

Figure 16.8
Schematic outline of vacuum filter

vacuum is released and the cake removed by scraper. Typical vacuum filter performance data are presented in Table 16.6.

Table 16.6 Typical vacuum filter performance data

Sludge	Chemical conditioning	Filter yield $(kg\ m^{-2}\ h^{-1})$	Cake solids (%)
Digested primary	Ferric chloride and lime	35	27
	Polymer	20–75	34–26
Mixed digested	Ferric chloride and lime	20–40	21.5
	Polymer	20	32–24

Two types of pressure filter are employed in sludge dewatering: the plate press and the belt press. An outline illustration of a plate press is shown in Figure 16.9. The plates are covered by a filter cloth and are surface-profiled to permit flow of the filtrate. Dewatering by the plate process involves filling, pressing and cake removal. Performance depends on sludge *SRF*, pressing time and final cake thickness. The pressing time required to achieve a given solids concentration is very

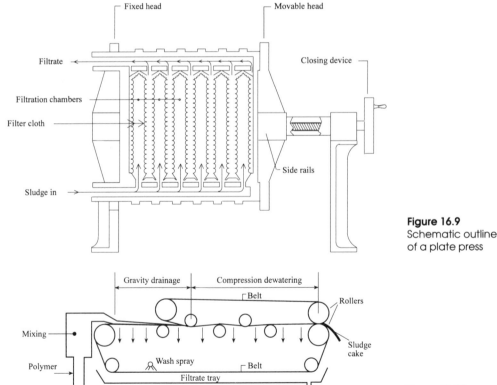

Figure 16.9
Schematic outline
of a plate press

Figure 16.10
Schematic outline
of a belt press

Figure 16.10a
Filter belt
press.
(Reproduced
courtesy of
Solids
Technology
Ltd, Dublin)

much influenced by the initial solids concentration (Gale, 1971); since all the expelled water has to be forced through the forming cake, the higher the initial solids concentration, the lower will be the amount of water to be discharged in this manner as filtrate.

A schematic outline of the belt process is shown in Figure 16.10. Preconditioned sludge is fed on to the horizontal drainage section of the unit, which removes up to 50% of the drainage water. The sludge layer is then sandwiched between the carrier belt and a cover belt and is subjected to a varying compression by the roller system. Only readily drainable sludges can be effectively dewatered by the belt press, which means that conditioning is almost invariably required. The dewatered cake solids concentration may vary in the range 12–25% by weight.

Centrifugation is a widely used industrial solids-separation process. It is also extensively used for sludge thickening and dewatering where solid bowl centrifuges are employed. Performance is influenced by sludge particle size and specific gravity. The sludge is fed into the rotating bowl where it is separated into a dewatered solids stream and a dilute centrate stream. Since the latter may contain a relatively high concentration of light solids, it must normally be returned for reprocessing. Sludge conditioning effects a reduction in the centrate solids concentration and an increase in the dewatered sludge solids concentration. The latter may lie in the range 10–25% by weight.

16.7   SLUDGE STABILIZATION

Many wastewater sludges possess high concentrations of biodegradable material, whose decomposition, in the absence of a continued supply of oxygen, is accompanied by the production of nuisance odour levels. Sludge can be stabilized by the biological conversion of its organic fraction to stable end products using aerobic or anaerobic digestion processes or by chemically inhibiting microbial activity in the sludge mass. The latter is usually effected by raising the sludge pH with lime.

16.7.1   Anaerobic digestion

The reader is referred to Chapter 15 for a detailed discussion of the theory and technology of anaerobic digestion processes. It is widely used for sludge stabilization and the treatment of high strength organic wastes. It can effect a 40–60% reduction in organic solids, improve dewaterability, reduce odour potential and produce methane as a useful byproduct. As shown in Figure 15.2, the rate of anaerobic decomposition is strongly influenced by temperature. Digestion proceeds rather slowly between 7°C and 16°C. Mesophilic (16–38°C) digestion is a more rapid process with optimal performance in the temperature range 30–35°C. Thermophilic digestion (45–65°C) is a still faster process and has an optimum around 55°C. The influence of temperature on digestion time for primary sewage sludge is shown in Figure 16.11.

In addition to being slow-growing, methane bacteria are sensitive to inhibition. Organic over-loading, leading to the accumulation of volatile acids and therefore a reduction in pH, may inhibit methane production. It is found that digestion proceeds satisfactorily in the pH range 6.6–7.6. Certain toxic components of wastewater sludges, including anionic detergents, chlorinated hydrocarbons, heavy metals as well as light metal cations at high concentrations (Swanwick, 1971), also inhi-

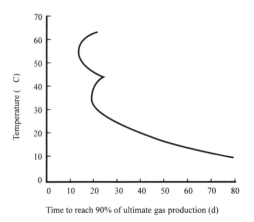

Figure 16.11 Influence of temperature on the digestion of primary sewage sludge. (Reprinted from Fair *etal*, 1968 by permission of John Wiley & Sons, Inc.)

Figure 16.12
Wastewater
treatment sludge
energy budget
(quantities relate
to 1 PE)

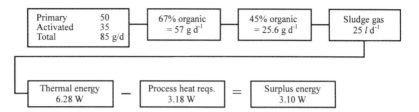

bit methane bacteria. Mesophilic digestion is commonly adopted as the primary mode of sludge stabilisation at medium- and large-sized works. Tanks are designed with a detention time in the range of 15–25 d with a mean suspended solids concentration in the range 3–6% dry matter. The process may be carried out in a single unit or in two stages, the second stage being used for settling and gas collection. Tanks are usually cylindrical with conical hopper bases and fixed or floating roofs, and are thermally insulated.

Adequate mixing is essential to ensure a uniform substrate concentration and effective solids degasification. Methods of mixing include mechanical stirring, sludge recirculation and gas recirculation. The heat required for the process is derived from combustion of the sludge gas produced. The quantity of gas produced generally varies in the range 0.9–1.4 m^3 per kg of volatile solids removed. Its composition is typically 65–70% of methane and 30–35% carbon dioxide with traces of hydrogen sulphide. As shown in Figure 16.12, the process, when applied to domestic sewage sludge, is potentially capable of rendering the treatment system self-sufficient in energy with some to spare.

16.7.2 Aerobic digestion

Aerobic stabilization refers to the separate aeration of primary, secondary or mixed sewage sludges resulting in the microbial oxidation of organic material. The primary oxidized end products are carbon dioxide and water. The process requires the input of oxygen through an aeration system. The amount of oxygen consumed depends on the volatile solids composition and has been found (Kambhu and Andrews, 1969) to vary from 1.2 to 1.5 kg oxygen per kg of volatile solids oxidized. The process is an exothermic one, the heat of reaction being approximately equal to the heat of combustion of the volatile solids (Fair and Moore, 1932).

The rate at which aerobic digestion proceeds is influenced by many factors, including the concentrations of microorganisms, biodegradable solids and inorganic nutrients, temperature, mixing intensity, etc. It has been found that conversion rates increase by about 3–7% per °C rise in temperature. Thus, aerobic digestion proceeds very rapidly in the thermophilic temperature range (45–65°C) where the evolution of heat can be sufficient to make the process thermally self-sustaining (Matsch and Drnevich, 1977).

Table 16.7 Aerobic digestion design parameters

Parameter	Value	Remarks
Hydraulic detention	15–20 d	Waste activated sludge alone.
	20–25 d	Primary + waste activated sludge
Air requirements Diffuser system	0.02–0.035 $m^3 \, min^{-1} \, m^{-3}$	Enough to keep the solids in suspension and maintain a D.O. between 1–2 mg l^{-1}. Waste activated sludge alone.
Mechanical system	0.06 $m^3 \, min^{-1} \, m^{-3}$ 26–33 Wm^3	Primary and waste activated sludge. This level is governed by mixing requirements. Most mechanical aerators in aerobic digesters require bottom mixers for solids concentration greater than 8000 mg l^{-1}, especially if deep tanks (> 4 m) are used.
Temperature	15°C	If sludge temperatures are lower than 15°C, then additional detention time should be provided so that stabilization will occur at the lower biological reaction rates.
Volatile solids reduction	40–50%	—
Tank design		Aerobic digestion tanks are open and generally require no special heat transfer equipment or insulation. For small treatment systems the tank design should be flexible enough so that the digester tank can also act as a sludge thickening unit. If thickening is to be utilized in the aeration tank, then sock-type diffusers should be used to minimize clogging.

Source: Weston, (1971).

The process, as generally employed, operates at ambient temperatures. The design is usually based on the single parameter of solids residence time which is normally in the range 10–20 d. Recommendations in the United States (Weston, Inc. 1971) for aerobic digester design are summarized in Table 16.7.

16.7.3 Chemical stabilization

Lime effects stabilization by raising the sludge pH to a level that inhibits all microbial activity. It would appear (Vesilind, 1975) that a sustained pH value in the region 11.0–12.2 achieves this objective, including the destruction of pathogenic organisms. The addition of lime to primary sludge has been found to reduce its specific resistance to filtration (Farrell *et al.*, 1974). The method, however, does not achieve a permanent stabilization because during long-term storage in lagoons, the pH gradually drops to a level which permits a resumption of microbial activity and so eventually producing nuisance conditions.

16.8 SLUDGE DISPOSAL

The selection of a specific disposal destination for sludge is one of the most important decisions to be made in wastewater treatment system design, particularly for larger works. The choice rests between disposal to agriculture, landfill or incineration with landfill disposal of the ash residue. It is influenced by the quantity and composition of the sludge and the location of the works relative to potential disposal sites. The extent of processing prior to disposal to any of these sites is determined by technical, economic and environmental considerations. Economic and environmental criteria are particularly important determinants. Sludge processing and disposal may account for over 50% of both the capital investment and operational costs of conventional activated sludge-based wastewater treatment systems. Environmental protection considerations impose constraints on disposal and exclude disposal options that are likely to cause environmental damage.

When sludge is used in agriculture it should be processed in such a manner as to retain the desired nutrients. Where sludge is to be transported in liquid form, considerable economy can usually be achieved by the use of a thickening processes to reduce its water content. Disposal to agriculture in thickened slurry form would appear to be an attractive solution, where it is economically and environmentally feasible. This method is, of course, climate-dependent and requires the provision of on-site storage to cater for those periods of the year when application to land is not possible.

REFERENCES

Albrecht, A. E. (1972) *J. Am. Water Works Assoc.*, **64**, 46–52.

Baskerville, R. C. and Gale, R. S. (1968) *Water Pollut. Control*, **67**, 233.

Baskerville, R. C., Bruce, A. M. and Day, M. C. (1978) *Filtr. Sep.*, **15**, 445–454.

Bratby, J. and Marais, G. R. (1977) *Publ. WP-20–4*, Washington, D.C.

Dick, R. I. (1972) *Filtr. Sep.*, **9**, 177–183.

Fair, G. M. and Moore, E. W. (1932) *Sewage Works J.*, **4**, 242, 428, 589, 728.

Farrell, J. B., Smith, J. E., Hathaway, S. W. and Dean, R. B. (1974) *J. Water Pollut. Control Fed.*, **46**, 113–122.

Fitch, B. (1975) *Filtr. Sep.*, **12**, 355–359.

Gale, R. S. (1971) *Proc. Fourth Public Health Eng. Conf.*, Loughborough, 33–50.

Kambhu, K. and Andrews, J. F. (1969) *J. Water Pollut. Control Fed.*, **41**, R127–R141.

Kalbskopf, K. H. (1972) *Water Res.*, **6**, 499–502.

Lockyear, C. F. (1977) *Tech. Rep. TR39*, Water Research Centre, Stevenage, England.

Matsch, L. C. and Drnevich, R. F. (1977) *J. Water Pollut. Control Fed.*, **49**, 296.

Popel, F. (1963) *Sludge Digestion and Disposal*, Technische Hochschule, Stuttgart.

Sarfert, F. (1976) *Prog. Water Technol.*, **8**, 353–356.

Swanwick, J. D. (1971) *Proc. Fourth Public Health Eng. Conf.*, Loughborough.

Swanwick, J. D. (1972) *Second European Sewage and Refuse Symposium*, Munich.

Swanwick, J. D. and Baskerville, R. C. (1966) *Proc. Inst. Sewage Purif.*, **65**, 153.

Van Gils, H. W. (1964) *Bacteriology of Activated Sludge*, Research Institute for Public Health Engineering, Delft, Netherlands.

Vesilind, P. A. (1975). *Treatment and Disposal of Wastewater Sludges*, Ann Arbor Science, Ann Arbor, Michigan.

Weston, R. F., Inc. (1971) *Upgrading Existing Wastewater Treatment Plants*, EPA Technology Transfer Contact 14–12–933.

White, J. B. (1970) *The Design of Sewers and Sewage Treatment Works*, Edward Arnold, London, 180.

WPCF (1969) Sludge Dewatering Manual of Practice No. 20.

Related reading

Bruce, A. M., Colin, F. and Newman, P. J., eds (1989) *Treatment of Sewage Sludge, Thermophilic Aerobic Digestation and Processing Requirements for Landfilling* Elsevier Applied Science Publishers, London.

Bruce, A. M., L'Hermite, P. and Newman, P. J., eds (1986) *New Developments in Processing of Sludges and Slurries*, Elsevier Applied Science Publishers, London.

Bruce, A. M., ed. (1984) *Sewage Sludge Stabilisation and Disinfection*, Ellis Horwood Ltd, Chichester, UK.

Casey, T. J., L'Hermite, P. and Newman, P. J. (1984) *Methods of Characterisation of Sewage Sludge*, D. Reidel Publishing Company, Dordrecht.

Cheremisinoff, P. N. (1994) *Sludge Management and Disposal*, Prentice-Hall Inc., New Jersey.

Vesilind, P. A., Hartman, P. A. and Skine, E. T. (1986) *Sludge Management and Disposal*, Lewis Publishers.

17

Disinfection

17.1 INTRODUCTION

The purpose of disinfection of water or wastewater is to kill or remove pathogenic (disease-causing) organisms. The groups of organisms implicated in the transmision of water-borne diseases include viruses, bacteria, protozoa and helminths (worms). Disinfection is a vital process in the production of drinking water and in the treatment of swimming pool waters. It is used to a much lesser extent in the treatment of wastewaters, where it may be applied, for example, to protect receiving waters used for bathing purposes.

The disinfection processes of significance in water and wastewater treatment can be classified into three categories:

(1) Chemical disinfectants, the more important of which are chlorine and ozone; other chemical disinfectants include iodine, phenolic compounds and silver nitrate.

(2) Physical disinfectants, such as heat and high-frequency (UV) electromagnetic radiation.

(3) Physical separation processes, such as centrifugation and membrane filtration.

17.2 DISINFECTION BY CHEMICAL AGENTS

In 1908 Chick showed that the rate of reduction of bacterial numbers by a disinfectant in a given contact time is proportional to the number of survivors at that time, i.e.

$$\frac{\mathrm{d}N}{\mathrm{d}t} = -KN \tag{17.1}$$

where N represents the number of living microorganisms, t denotes the contact time and K is the rate constant or coefficient of lethality.

Equation (17.1) is generally referred to as 'Chick's Law'. Integration of equation (17.1) yields:

$$\ln \left(\frac{N}{N_0}\right) = -Kt \qquad (17.2)$$

where N_0 is the initial number of microorganisms and N is the surviving number after a disinfectant contact time t.

It emerges that if the lethal action of a particular disinfectant on a microbial population obeys Chick's law and if the coefficient of lethality is independent of N and t, then the plot of the logarithm of the fraction or number of microorganisms surviving against time will be linear. However, it should be noted that the number of disinfectant–microorganism combinations exhibiting exact conformity with Chick's Law is rather limited, since the coefficient of lethality is often time-dependent.

A time-dependent rate of kill may be expressed in the form:

$$\frac{dN}{dt} = -KNt \qquad (17.3)$$

for which the integrated form is

$$\ln \left(\frac{N}{N_0}\right) = -K\frac{t^2}{2} \qquad (17.4)$$

It has been found that the biocidal action of the halogens and other oxidizing disinfectants rarely conforms to Chick's Law, being more accurately represented by a time-dependent model, such as defined by equation (17.3).

17.3 DISINFECTION BY ULTRA VIOLET LIGHT

Ultraviolet light kills the vegetative and spore cells of microorganisms. It is also lethal to viruses. The lethality of ultraviolet light varies greatly with its wavelength. The bactericidal wavelengths extend from 200 nm to 295 nm, with a maximum effect at about 254 nm.

The rate of devitalization of a population of microorganisms due to irradiation with ultraviolet light obeys Chick's Law. The coefficient of lethality, K, is a function of the radiation intensity:

$$K = K'I^n \qquad (17.5)$$

where K' is a constant, I is the radiation intensity, and n is a constant, approximating to unity in value.

The intensity of radiation is reduced as it passes through an absorbing medium. This reduction can be represented as follows:

$$\frac{I}{I_i} = e^{-ax} \qquad (17.6)$$

where I_i is the intensity of the incident radiation at the medium surface, I is the intensity after it has travelled a distance x into the absorbing medium (x is referred to as the light path) and a is a constant, which is sometimes called the 'linear absorption coefficient'.

The fraction I/I_i is known as the 'transmittance' or 'opacity', denoted by T; hence, equation (17.6) may be written in the form

$$\log \frac{1}{T} = 0.4343\ ax \tag{17.7}$$

Log $(1/T)$ is known as the 'optical density' or 'absorbance' or 'extinction'. $[0.4343\ a]$ is called the 'extinction coefficient'.

It is at once apparent that the plot of extinction against light path will be linear and that the slope of this plot yields the extinction coefficient.

Pure water absorbs ultraviolet light. Natural and waste waters may well contain a variety of molecular and ionic species that absorb ultraviolet light. The absorption coefficient is a function of the concentration of the absorbing species. The relationship between transmittance and concentration was shown by Beer, in 1852, to take the form (for a single absorbing species)

$$T = e^{-a'c} \tag{17.8}$$

where a' is a constant and c is the concentration of the absorbing species.

Combining equations (17.7) and (17.8) gives the 'Beer–Lambert Law', which expresses the relation between transmittance, light path and concentration of the absorbing material:

$$T = e^{-a''cx} \tag{17.9}$$

where a'' is a constant.

It follows that

$$\log \frac{1}{T} = 0.4343\ a''cx \tag{17.10}$$

$0.4343\ a''$ is called the 'specific extinction coefficient'.

It is important to note that the magnitudes of a, a' and a'' are functions of the radiation wavelength and may well be significantly dependent on other factors, such as temperature.

For the purposes of water disinfection, ultraviolet light emanates from a source and penetrates the water bulk. Hence, the rate of kill varies throughout the water depth. An average coefficient of lethality can be computed by means of the relation

$$K_{avg} = \frac{\displaystyle\int_0^x K\ \mathrm{d}x}{x} \tag{17.11}$$

Combining equations (17.5), (17.6) and (17.11) and assuming n to have unit value, the average coefficient of lethality is found to be

$$K_{avg} = \frac{K'I_i}{x} \left[\frac{1}{a} - \frac{e^{-ax}}{a} \right] \qquad (17.12)$$

This average coefficient of lethality can then be incorporated into equations (17.2)–(17.4), inclusive.

Light-scattering particles will also reduce the intensity of the electro-magnetic radiation. However, the effect of such particles on the light intensity does not lend itself to a generalized mathematical treatment since their ability to scatter light is a function of many variables such as the size, shape and the chemical nature of the particle. A mathematical treatment is further complicated by the fact that light-scattering parti-cles may also absorb light.

17.4   DISINFECTION BY HEAT

Disinfection by heat also obeys Chick's Law. The rate constant ana-logous to the coefficient of lethality is a function of the thermal resis-tance of the target organism. It is important to note that the thermal resistance of the vegetative and spore cells of the same organism may well differ considerably, the spore cells being far more resistant to thermal devitalization. The situation is further complicated by the fact that the rate constant, being temperature-dependent, will vary with time since the disinfection procedure will involve increasing the temperature, holding temperature at a fixed level and then cooling the system.

Milk pasteurization is an example of disinfection by heat. The tem-perature range used is usually 60–80°C. The required exposure time at 60°C is about 20 min; it decreases with an increase in temperature.

17.5   DISINFECTION BY SEPARATION PROCESSES

Microrganisms are sometimes removed from fluids, both gases and liquids, by filtration using very fine filters. They can also be removed from liquids by the application of great centrifugal forces.

A significant reduction in the bacterial content of natural and waste waters is accomplished by processes where the primary function is to remove coarse organic debris and suspended particles. The use of coarse and fine screens, grit chambers, slow and rapid sand filters and plain sedimentation is generally accompanied by a reduction in the bacterial population of the water. Chemical precipitation processes also reduce the bacterial population.

17.6   DISINFECTION WITH CHLORINE (white, 1986)

Chlorine (Cl_2) is a greenish-yellow gas which can be readily compressed to form a clear amber-coloured liquid. It vaporizes readily at normal atmospheric pressure and temperature and has a characteristic irritating, penetrating and pungent odour. It has a critical temperature (the temperature above which it can exist only as a gas) of 144°C and a critical pressure (the vapour pressure at critical temperature) of 7.712 $MN\,m^{-2}$ absolute. On gasification at 0°C and 1 atm, its volume expands by a factor of 457.6. Liquid chlorine has a density of 1468 kg m^{-3}, while gaseous chlorine has a density of 3.21 kg m^{-3} at 1 atm and 0°C (i.e. about 2.48 times that of air). It is highly soluble in water (*ca.* 1% by weight at 10°C), as may be seen from the solubility data given in Table 1.6.

17.6.1 Chemical reactions

When chlorine gas is added to water, the following reactions (Snoeyink and Jenkins, 1980) occur:

$$Cl_2 + H_2O \Leftrightarrow HOCl + HCl \Leftrightarrow 2H^+ + Cl^- + OCl^- \qquad (17.13)$$

where HOCl is hypochlorous acid, HCl is hydrochloric acid and OCl is the hypochlorite ion.

At ordinary water temperatures, this reaction reaches equilibrium within a few seconds. In dilute solutions and pH levels above 4, the reaction is pushed towards the right, resulting in very little chlorine remaining as chlorine gas in the water. The corresponding equilibrium expression is

$$\frac{[HOCl][H^+][Cl^-]}{[Cl_2]} = K_s = 4.5 \times 10^{-4} \text{ at } 20°C \qquad (17.14)$$

where the square brackets denote molar concentration.

Hypochlorous acid is a weak acid, the degree of dissociation or ionization being given by

$$\frac{[H^+][OCl^-]}{[HOCl]} = K_i = 2.5 \times 10^{-8} \text{ at } 20°C \qquad (17.15)$$

K_i is known as the 'ionization' or 'dissociation' constant.

Dissociation of hypochlorous acid reaches equilibrium virtually instantaneously. It is at once apparent from equation (17.15) that the degree of dissociation is pH-dependent. It should be noted that K_s and K_i are temperature-dependent.

Chlorine gas and hypochlorous acid are potent germicidal agents. The hypochlorite ion also exerts a lethal effect but is not as potent as the former species. Since very little chlorine remains as chlorine gas in water, the principal disinfecting action of chlorine is associated with hypochlorous acid.

Chlorine existing in water as hypochlorous acid and hypochlorite ions is defined as 'free available chlorine'.

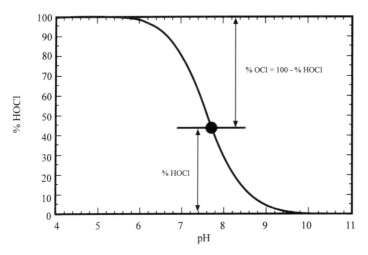

Figure 17.1
Dissociation of
hypochlorous acid
as a function of pH

The relative distribution of hypochlorous acid and the hypochlorite ion is very important since the former is about 40–80 times more efficient as a disinfecting agent than the latter. Figure 17.1 shows their relative amounts as a function of pH, as defined by the equilibrium equation (17.15).

Hypochlorous acid is also produced when hypochlorites such as calcium and sodium hypochlorite are dissolved in water:

calcium hypochlorite: $Ca(OCl)_2 \Leftrightarrow Ca^{2+} + 2OCl^-$ (17.16)
sodium hypochlorite: $NaOCl \Leftrightarrow Na^+ + OCl^-$ (17.17)

A certain fraction of the hypochlorite ions will combine with hydrogen ions to produce hypochlorous acid. Equilibrium will be established in accordance with equation (17.15). Thus, the same equilibria are established in water regardless of whether elemental chlorine or hypochlorite is employed. The important distinction is the resultant pH and hence the relative amounts of hypochlorous acid and hypochlorite ions existing at equilibrium. Chlorine tends to reduce the initial pH, whereas hypochlorite salts tend to raise it.

When chlorine is added to water containing natural or added ammonia, the ammonia reacts with hypochlorous acid to form various chloramines:

$$NH_3 + HOCl \Leftrightarrow NH_2Cl + H_2O$$
$$\text{monochloramine}$$ (17.18)

$$NH_2Cl + HOCl \Leftrightarrow NHCl_2 + H_2O$$
$$\text{dichloramine}$$ (17.19)

$$NHCl_2 + HOCl \Leftrightarrow NCl_3 + H_2O$$
$$\text{trichloramine or}$$ (17.20)
$$\text{nitrogen trichloride}$$

The distribution of the reaction products is governed by pH, temperature, time and the initial $Cl_2:NH_3$ ratio. The situation is complicated by the fact that as additional amounts of chlorine are applied, the chloramines are eventually destroyed. Thus, the chloramines will exist while ammonia is in excess but are destroyed when chlorine is in excess. The following equations suggest the reactions that might occur in the destruction of the chloramines:

$$HOCl + 2NH_2Cl \Leftrightarrow N_2 + 3HCl + H_2O \qquad (17.21)$$

$$4NH_2Cl + 3Cl_2 + H_2O \Leftrightarrow N_2 + N_2O + 10HCl \qquad (17.22)$$
$$\text{nitrous}$$
$$\text{oxide}$$

$$2NHCl_2 + H_2O \Leftrightarrow N_2O + 4HCl \qquad (17.23)$$

The chloramines are effective disinfecting agents but are not as potent as hypochlorous acid.

Chlorine also reacts with organic nitrogen to form compounds analogous to the products formed when it combines with free ammonia.

Chlorine which is in chemical combination with ammonia or organic nitrogen is defined as 'combined available chlorine'.

Free chlorine readily oxidizes reduced inorganic substances such as hydrogen sulphide, nitrites, iron and manganese (in the ferrous and manganous forms):

$$H_2S + 4HOCl \Leftrightarrow H_2SO_4 + 4HCl \qquad (17.24)$$

$$2Fe^{2+} + HOCl + H^+ \Leftrightarrow 2Fe^{3+} + Cl^- + H_2O \qquad (17.25)$$

$$2Mn^+ + HOCl + H^+ \Leftrightarrow 2Mn^{2+} + Cl^- + H_2O \qquad (17.26)$$

$$NO_2^- + HOCl \Leftrightarrow NO_3^- + HCl \qquad (17.27)$$

Chloramines do not enter vigorously into these oxidation reactions.

The various substances that react with chlorine as it is added to water are said to exert a 'chlorine demand'. The above inorganic species, which readily react with chlorine, are said to constitute the 'immediate chlorine demand'. A plot of the chlorine residual against the amount of chlorine added is presented in Figure 17.2.

'Combined residual chlorination' practice involves the application of chlorine to water to produce, with natural or added ammonia, a combined available chlorine residual with the objective of maintaining a chlorine residual through part or all of a water treatment plant or distribution system.

'Free residual chlorination' practice involves the application of chlorine to water to produce, either directly or through the destruction of ammonia, a free available chlorine residual and to maintain that residual through part or all of a water treatment plant or distribution system.

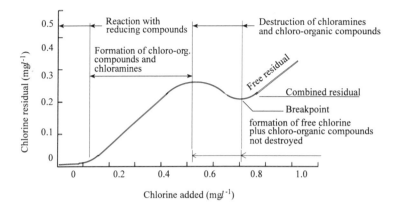

Figure 17.2
Influence of
chlorine dose on
chlorine residual

17.6.2 Chlorine compounds

The chlorine-releasing compounds used in water/wastewater treatment plants include chlorine gas, calcium hypochlorite, sodium hypochlorite and chlorine dioxide. Calcium and sodium hypochlorites are used in very small treatment plants, such as package plants, where simplicity and safety are more important than cost.

Calcium hypochlorite is available commercially in either a 'dry' (solid) or a 'wet' (dissolved) form. Because the solid material is relatively stable, under proper storage conditions it is often favoured over the wet form. Because hypochlorite is a strong oxidizing agent it should be stored in a cool, dry location in corrosion-resistant containers and away from other chemicals.

Sodium hypochlorite is available as a weak (usually 3%) aqueous solution. Consequently, transportation costs may limit its application. The solution decomposes more readily at high concentrations and is affected by exposure to light and heat. It must be stored in corrosion-resistant containers and in a cool place.

Chlorine gas is supplied in liquefied form under high pressure. Chlorine gas is an extremely corrosive and poisonous gas. Being heavier than air, adequate exhaust ventilation at floor level must be provided. Chlorine storage and chlorination equipment rooms should be walled off from the rest of the plant and should be accessible only from out-of-doors. It is recommended that a fixed glass viewing window should be included in an inside wall. Fan controls and gas masks should be located at the entrance to the room. Dry chlorine gas and liquid chlorine can be handled in black wrought iron piping but chlorine solution is very corrosive and should be handled in rubber-lined or tough plastic piping with diffusers of hard rubber. The gas cylinders in use should be mounted on platform scales set flush with the floor so that the loss of weight can be used as a positive record of chlorine dosage.

Chlorine dioxide (ClO_2) is used in place of gaseous chlorine in situations where the reactions between the latter and certain trace organic substances, such as phenolic compounds, give rise to bad taste or odour in drinking water. Chlorine dioxide is a reddish-yellow gas at ambient temperature ($>10°C$) and is highly soluble in water (see Table 1.7). It is produced by reaction between chlorine and sodium chlorite:

$$2NaClO_2 + Cl_2 \Leftrightarrow 2ClO_2 + 2NaCl \qquad (17.28)$$

17.6.3 Trihalomethane formation

As shown on Figure 17.2, when chlorine is added to water it reacts with dissolved organic compounds to form a variety of chloro-organic compounds. As noted in section 7.4 of Chapter 7, organohalogen compounds as a group are listed as dangerous to living organisms. Of particular concern in the context of drinking water disinfection by chlorination are the trihalomethanes (THMs), which are formed from the reaction between *free* chlorine and naturally occurring humic substances in water. The THMs have the general formula CHX_3, where X can be Cl, Br or I. Chloroform ($CHCl_3$) is of particular concern in drinking water because it is a suspected carcinogen. The brominated species which may be formed include dichlorobromomethane ($CHCl_2Br$), dibromochloromethane ($CHClBr_2$) and bromoform ($CHBr_3$). The total THM concentration (TTHM) in chlorinated drinking water is a function of the organic precursor concentration and the chlorine dose used, as illustrated in Figure 17.3. This data set is taken from a study of THM formation in Irish drinking water supplies (Chua, 1996), being the highest and lowest values measured. It will be noted that the water from plant A has a higher TOC and, hence, a higher chlorine demand per dose than plant B. These factors combine to produce a much higher TTHM generation in the water from plant A than in the water from plant B.

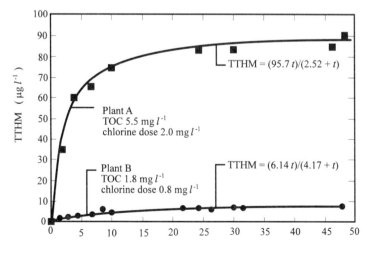

Figure 17.3
Formation of trihalomethanes in drinking water

As may be noted from Figure 17.3, the THM concentration continues to increase over a relatively long period following chlorination, following a decreasing rate growth curve, which may be represented by a hyperbolic function of the form:

$$\text{TTHM}_t = \text{TTHM}_{\max}\left(\frac{t}{t_{50} + t}\right) \tag{17.29}$$

where t_{50} is the time (h) required to reach 50% of $\text{TTHM}_{\max}$. As indicated in Figure 17.3, t_{50} was 2.52 h for plant A water (high TTHM water) and 4.17 h for plant B water (low TTHM water).

THM formation can be greatly reduced by:

(a) the removal of the organic precursors (colour-causing humic substances) by chemical coagulation prior to chlorination;

(b) a reduction in the chlorine dose; and

(c) the addition of ammonia to form chloramines in place of free chlorine (so-called chloramination).

17.6.4 Chlorination practice in drinking water treatment

Chlorination practice in drinking water production is governed by the requirement to produce a finished water that meets the prevailing drinking water standards both in respect of microbiological quality and chloro-organic residuals. The microbiological quality requirement is generally deemed to be met if testing shows zero concentrations of total coliforms, fecal coliforms, fecal streptococci and sulphate-reducing clostridia (WHO, 1992: EU Drinking Water Directive, 80/778/EEC).

The disinfection effectiveness of chlorination depends on the chlorine dose and the chlorine contact time. As shown on Figure 17.2, chlorine reacts with organics and inorganics which exert an immediate chlorine demand, thus reducing the residual free chlorine (RFC) available for disinfection. The product of the RFC concentration and contact time gives the process CT-value, which is probably the best measure of the disinfection potential of a chlorination process. Figure 17.4 illustrates a typical RFC/time decay curve—the CT-value being the area below the curve. The CT-value can be approximated with reasonable accuracy (Chua, 1996) by assuming an exponential decay rate for the RFC and using the 2-h RFC value to define the decay rate constant:

$$C_t = C_0 e^{-kt} \tag{17.30}$$

Hence

$$k = 0.5\ln\left(\frac{C_0}{C_2}\right)h^{-1} = \frac{\ln(C_0/C_2)}{120}\text{min}^{-1}$$

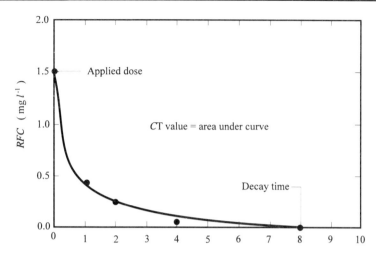

Figure 17.4
Typical chlorine
decay curve

where k is the decay rate constant (d^{-1}), C_0 is the initial RFC-value and C_2 is the RFC-value after 2 h contact time. The CT-value is found by integration over time:

$$CT = \int_0^\infty C_0 e^{-kt} dt = \frac{C_0}{k}$$

Hence

$$CT = \frac{120C_0}{\ln\left(\dfrac{C_0}{C_2}\right)} \, \text{mg l}^{-1} \, \text{min} \tag{17.31}$$

The WHO Drinking Water Guidelines (1992) recommend a minimum RFC of 0.5 mg l^{-1} after 30 min contact time at pH 7 for waters with a turbidity of less than 1 NTU. It should be noted that the use of a standard chlorination rate defined in this way does not result in a standard CT-value. In a series of tests using the recommended WHO guideline chlorination on 12 different drinking waters, which had been pre-treated by chemical coagulation/filtration processes, the CT-values were found (Chua, 1996) to vary in the range 48–136 mg l^{-1} min. Because of the variation in the initial decay rate for different waters, it is preferable to regulate chlorination by specifying the RFC after a longer contact period than 30 min, say 2 h. Equation (17.31) may then be used to calculate the corresponding CT-value. For waters that have been pre-treated by chemical coagulation/filtration, a CT-value of 50 mg l^{-1} min should be adequate for drinking water disinfection.

17.7 DISINFECTION WITH OZONE

Ozone (O_3) is a powerful oxidizing agent. It is used for drinking water disinfection and for air de-odourization in wastewater pumping stations and in sludge-handling facilities.

It is a blue gas having a characteristic, but not unpleasant, odour. It is formed from oxygen (O_2) by passing a silent electric discharge through air. Care must be taken to avoid the formation of an electric arc which could cause deterioration of the electrodes and produce nitrous vapours. This is ensured by using dry air and by covering one or both of the electrodes with a glass dielectric. The voltage applied between the electrodes is generally in the range 8–15 kV. At a frequency of 50 cycles per second, the production of ozone per square metre of discharge area is of the order of 14 mg s^{-1} (i.e. 50 g h^{-1}) and the power consumed is 1 kW. This output is doubled if a frequency of 500 cycles per second is used, an important factor in large installations. The ozone output decreases with increasing temperature. Consequently, the electrodes must be kept as cool as possible; this is generally accomplished by circulating water or oil.

The concentration of ozone in the ozonized air is, on average, from 5 to 10 g m^{-3}. To sterilise 1 m^3 of reasonably clear water, from 0.5 to 1.0 g of ozone is needed. Greater quantities are required if the organic content of the water is significant. Neutral salts and hydroxyl ions accelerate its decomposition , which in pure aqueous solution occurs as:

$$O_3 + H_2O \rightarrow HO_3^+ + OH^-$$
$$H_3O^+ + OH^- \rightarrow 2H_2O$$
$$O_3 + HO_2 \rightarrow HO + 2O_2$$
$$HO + HO_2 \rightarrow H_2O + O_2$$

The free radicals HO_2 and HO retain great oxidizing potential, and they are reactive with many common impurities such as metal salts and organic matter.

Since ozone has a low solubility in water (see Table 1.7) and its concentration (and therefore partial pressure) is low in the air or oxygen stream in which it is carried, an efficient gas–water contacting system must be used for its dissolution. The ozonized air or oxygen stream is

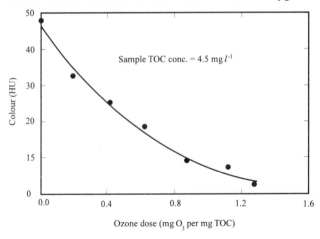

Figure 17.5
Colour reduction by ozone. *Source:* Casey and Chua (1995)

introduced into the water through a submerged diffuser system or by means of a packed bed gas–water dissolution system, using materials that are chemically non-reactive with ozone.

Because of its strong oxidizing capacity, ozone can be used to remove natural colour from water by a bleaching action. An example of the bleaching action of ozone on water colour is given in Figure 17.5.

REFERENCES

Casey, T. J. and Chua, H. K. (1995) Trihalomethanes in drinking water, paper presented to the Water & Environmental Section, Institution of Engineers of Ireland, Dublin

Chua, H. K. (1996) *Trihalomethane Precursor Control in Drinking Water Treatment*, PhD Thesis, Department of Civil Engineering, University College Dublin.

Snoeyink, V. L. and Jenkins, D. (1980) *Water Chemistry*, John Wiley & Sons, New York.

WHO (1992) Revision of WHO Guidelines for Drinking Water Quality, WHO, Geneva.

Related reading

AWWA (1986) *Water Chlorination, Principles and Practice*, American Waterworks Association, Denver, CO 80235, USA.

Jolley, R. L. ed. (1990) *Water Chlorination, Chemistry, Environmental Impact and Health Effects*, Vol. 6, Lewis Publications.

White, G. C. (1986) *Handbook of Chlorination*, Van Nostrand Reinhold, New York.

Appendix A

Listings of the computer programs FLUPROPS.BAS, GASSOL.BAS, DECANT.BAS and OCFIT.BAS

Computer programs in source code GWBASIC providing databases and computational techniques for designing and analysing water and wastewater treatment systems, are available in executable form on disk, free of charge, from:
Customer Services
John Wiley & Sons, Ltd., Distribution Centre, Southern Cross Trading Estate
Shripney Road, Bognor Regis, West Sussex PO22 9SA, UK.
Phone:+44 1243 843 294

ISBN 0471-97132-4 disk   Free of charge

LISTING OF THE PROGRAM FLUPROPS.BAS:

```
 10 REM PROGRAM FLUPROPS
 20 REM THIS PROGRAM PROVIDES A DATA BASE OF
    FLUID PROPERTIES
 30 DIM DENS(30), VISC(30), STEN(30), SVP(30)
 40 FOR I=1 TO 21:READ DENS(I), VISC(I), STEN(I),
    SVP(I): NEXT I
 50 READ AICP, AICV, AIA, AIB, AIC, AID, AIS
 60 READ OXCP, OXCV, OXA, OXB, OXC, OXD, OXS
 70 READ NICP, NICV, NIA, NIB, NIC, NID, NIS
 80 READ MECP, MECV, MEA, MEB, MEC, MED, MES
 90 READ CDCP, CDCV, CDA, CDB, CDC, CDD, CDS
100 CLS:PRINT TAB(5) "PROGRAM FLUPROPS": PRINT
110 PRINT TAB(5) "TO ACCCESS WATER DATA, ENTER 1"
120 PRINT TAB(5) "TO ACCESS GASES DATA, ENTER 2"
130 PRINT: PRINT TAB(5) "ENTER 1 OR 2, AS
    APPROPRIATE"
140 INPUT SEL: IF SEL <1 OR SEL<2 THEN GOTO 130
150 ON SEL GOTO 170, 340
160 REM PHYSICAL PROPERTIES OF WATER
```

```
170 PRINT: INPUT "ENTER THE WATER TEMPERATURE
    (deg C)"; T
180 I=INT (T/5)+1
190 DENS=DENS(I)+(T-(I-1)*5)*(DENS(I+1)-
    DENS(I))/5
200 VISC=VISC(I)+ T-(I-1)*5)*(VISC(I+1)-
    VISC(I))/5
210 STEN=STEN(I)+ T-(I-1)*5)*(STEN(I+1)-
    STEN(I))/5
220 SVP=SVP(I)+T-(I-1)*5)*(SVP(I+1)-SVP(I))/5
230 PRINT:PRINT "PHYSICAL PROPERTIES OF WATER
    AT"; T;
240 PRINT "deg C ARE AS FOLLOWS:": PRINT
250 PRINT "DENSITY (kg/m**3) ="; PRINT
260 PRINT "DYNAMIC VISCOSITY (Ns/m**2) =";
    VISC
270 PRINT "SURFACE TENSION (N/m) ="; STEN*.001
280 PRINT "SATURATION VAPOUR PRESSE
    (N/m**2)="; SVP
290 PRINT: PRINT "PRESS THE SPCAE BAR TO
    CONTINUE": Y$=INPUT$(1)
300 PRINT: INPUT "DO YOU WISH TO OBTAIN FURTHER
    DATA (Y/N)"; ANS$
310 IF ANS$="Y" THEN GOTO 100
320 IF ANS$<>"N" THEN GOTO 300 ELSE GOTO 1040
330 REM PHSYCIAL PROPERTIES OF GASES
340 PRINT TAB(20) "1. AIR"
350 PRINT TAB(20) "2. OXYGEN"
360 PRINT TAB(20) "3. NITROGEN"
370 PRINT TAB(20) "4. METHANE"
380 PRINT TAB(20) "5. CARBON DIOXIDE"
390 PRINT: INPUT "SELECT GAS BY BY TYPEING ITS
    NUMBER"; Z
400 PRINT: IF Z<1 OR Z>5 THEN GOTO 390
410 INPUT "INPUT GAS TEMPERATURE (deg C)"; T
420 INPUT "INPUT GAS ABSOLUTE PRESSURE
    (N/m**2); P
430 ON Z GOTO 440, 560, 680, 800, 920
440 REM AIR PROPERTIES
450 T=T+273.2: AIR=AICP-AICV: RHO=p/(AIR*T)
460 VISC=AIS*2.71828^(AIA*LOG(T)+(AIB/T)
    +(AIC/(T*T)+AID)
470 PRINT: PRINT "DENSITY OF AIR (kg/m**3) =";
    RHO
480 PRINT "DYNAMIC VISCOSITY OF AIR
    (Ns/m**2)=";VISC
490 PRINT "SPECIFIC HEAT AT CONSTANT PRESSURE
    (J/kg.K) ="; AICP
500 PRINT "SPECIFIC HEAT AT CONSTANT VOLUME
    (J/kg.K)="; AICV
```

```
510 PRINT "SPECIFIC GAS CONSTANT (J/kg.K) =";
    AIR
520 PRINT: PRINT "PRESS THE SPACE BAR TO
    CONTINUE": Y$=INPUT$(1)
530 PRINT: INPUT "DO YOU WISH TO OBTAIN FURTHER
    DATA (Y/N)"; ANS$
540 IF ANS$ = "Y" THEN GOTO 100
550 IF ANS$<> "N" THEN GOTO 530 ELSE GOTO 1040
560 REM OXYGEN PROPERTIES
570 T=T+273.2: OXR=OXCP-OXCV: RHO=P/(OXR*T)
580 VISC=OXS*2.71828^(OXA*LOG(T)+(OXB/
    T)=OXC/(T*T)+OXD)
590 PRINT: PRINT "DENSITY OF OXGYEN (kg/m**3)
    ="; RHO
600 PRINT "DYNAMIC VISCOSITY OF OXYGEN (Ns/
    m**2) ="; VISC
610 PRINT "SPECIFIC HEAT AT CONSTANT PRESSURE
    (J/kg.K) ="; OXCP
620 PRINT "SPECIFIC HEAT AT CONSTANT VOLUME (J/
    kg.K) ="; OXCV
630 PRINT "SPECIFIC GAS CONSTANT (J/kg.K) =";
    OXR
640 PRINT: PRINT "PRESS THE SPACE BAR TO
    CONTINUED": Y$=INPUT$(1)
650 PRINT: INPUT "DO YOU WISH TO OBTAIN FURTHER
    DATA (Y/N)"; ANS$
660 IF ANS$= "Y" THEN GOTO 100
670 IF ANS$<> "N" THEN GOTO 650 ELSE GOTO 1040
680 RTEM NITROGEN PROPERTIES
690 T=T+273.2: NIR=NICP-NICV: RHO=P/(NIR*T)
700 VISC=NIS*2.71828^(NIA*LOG(T)+(NIB/T)
    +NIC/(T*T)+NID)
710 PRINT: PRINT "DENSITY OF NITROGEN (kg/m**3)
    ="; RHO
720 PRINT "DYNAMIC VISCOSITY OF NITROGEN (Ns/
    m**2) ="; VISC
730 PRINT "SPECIFIC HEAT AT CONSTANT PRESSURE
    (J/kg.K) ="; NICP
740 PRINT "SPECIFIC HEAT AT CONSTANT VOLUME
    (J/kg.K). ="; NICV
750 PRINT "SPECIFIC GAS CONTANT (J/kg.K) =";
    NIR
760 PRINT: PRINT "PRESS THE SPACE BAR TO
    CONTINUE": Y$=INPUT$(1)
770 PRINT: INPUT "DO YOU WISH TO OBTAIN FURTHER
    DATA (Y/N)"; ANS$
780 IF ANS$= "Y" THEN GOTO 100
790 IF ANS$<> "N" THEN GOTO 770 ELSE GO TO 1040
800 REM METHANE PROPERTIES
810 T=T+273.2: MER=MECP-MECV: RHO=P/(MER*T)
```

```
820 VISC-MES*2.71828^(MEA*LOG(T)+(MEB/T)
    +MEC/(T*T)+MED)
830 PRINT: PRINT "DENSITY OF METHANE (kg/m**3)
    ="; RHO
840 PRINT "DYNAMIC VISCOSITY OF METHANE (Ns/
    m**2)="; VISC
850 PRINT "SPECIFIC HEAT AT CONSTANT PRESSURE
    (J/kg.K)="; MECP
860 PRINT "SPECIFIC HEAT AT CONSTANT VOLUME (J/
    kg.K).="; MECV

870 PRINT "SPECIFIC GAS CONTANT (J/kg.K)=";
    MER
880 PRINT: PRINT "PRESS THE SPACE BAR TO
    CONTINUE": Y$=INPUT$(1)
890 PRINT: INPUT "DO YOU WISH TO OBTAIN FURTHER
    DATA (Y/N) "; ANS$
900 IF ANS$="Y"THEN GOTO 100
910 IF ANS$<>"N"THEN GOTO 890 ELSE GO TO 1040
920 REM CARBON DIOXIDE PROPERTIES
930 T=T+273.2: CDR=CDCP-CDCV: RHO=P/(CDR*T)
940 VISC=CDS*2.71828^(CDA*LOG(T)+(CDB/
    T)+CDC/(T*T)+CDD)
950 PRINT: PRINT "DENSITY OF CARBON DIOXIDE
    (kg/m**3)="; RHO
960 PRINT "DYNAMIC VISCOSITY OF CARBON DIOXIDE
    (Ns/m**2)="; VISC
970 PRINT "SPECIFIC HEAT AT CONSTANT PRESSURE
    (J/kg.K)="; CDCP
980 PRINT "SPECIFIC HEAT AT CONSTANT VOLUME (J/
    kg.K). ="; CDCV
990 PRINT "SPECIFIC GAS CONTANT (J/kg.K) =";
    CDR
1000 PRINT: PRINT "PRESS THE SPACE BAR TO
     CONTINUE": Y$=INPUT$(1)
1010 PRINT: INPUT "DO YOU WISH TO OBTAIN FURTHER
     DATA (Y/N) "; ANS$
1020 IF ANS$="Y"THEN GOTO 100
1030 IF ANS$<>"N"THEN GOTO 1010 ELSE GO TO 1040
1040 END
1050 DATA 990.968, 1.787E-3, 75.6, 6.6107E3
1060 DATA 999.992, 1.519E-3, 74.9, 0.8721E3
1070 DATA 999.728, 1.307E-3, 74.22, 1.2277E3
1080 DATA 999.129, 1.139E-3, 73.49, 1.7049E3
1090 DATA 998.234, 1.002E-3, 72.75, 2.3378E3
1100 DATA 997.075, 0.8904E-3, 71.97, 3.1676E3
1110 DATA 995.678, 0.7975E-3, 71.18, 4.2433E3
1120 DATA 994.064, 0.7194E-3, 70.37, 5.6237E3
1130 DATA 992.247, 0.6529E-3, 69.56, 7.3774E3
1140 DATA 990.24, 0.5960E-3, 68.74, 9.5848E3
1150 DATA 998.07, 0.5468E-3, 67.91, 12.338E3
```

```
1160 DATA 985.73, 0.5040E-3, 67.05, 15.745E3
1170 DATA 983.24, 0.4665E-3, 66.18, 19.924E3
1180 DATA 980.59, 0.4335E-3, 65.29, 25.013E3
1190 DATA 977.81, 0.4042E-3, 64.40, 31.166E3
1200 DATA 974.89, 0.37813-3, 63.50, 38.553E3
1210 DATA 971.83, 0.3547E-3, 62.60, 47.364E3
1220 DATA 968.65, 0.3337E-3, 61.68, 57.808E3
1230 DATA 965.34, 0.3147E-3, 60.76, 70.112E3
1240 DATA 961.92, 0.2975E-3, 59.84, 84.528E3
1250 DATA 958.38, 0.2818E-3, 58.90, 101.325E3
1260 DATA 1005.0, 717.9, 0.63404, -45.638,380.87,
        -3.4505, 182.0E-7
1270 DATA 920.0, 657.1, 0.52662, -97.589, 2650.7,
        -2.6892, 203.2E-7
1280 DATA 1040.0, 742.9, 0.60097, -57.005,
        1029.1, -3.2322, 175. 7E-7
1290 DATA 2260.0, 1725.2, 0.54188, -127.57,
        4700.8, -2.6952, 109.3E-7
1300 DATA 876.0, 673.8, 0.44037, -288.4, 19312.0,
        -1.7418, 146.7E-7
```

LISTING OF THE PROGRAM GASSOL.BAS:

```
 10 REM PROGRAM GASSOL
 20 REM THIS PROGRAM PROVIDES DATA ON THE
    SOLUBILITY OF GASES IN WATER
 30 DIM SOL (12, 12)
 40 FOR I=1 TO 5
 50 FOR J=1 TO 11: READ SOL(I,J): NEXT J
 60 NEXT I
 70 FOR I=6 TO 11
 80 FOR J=1 TO 7: READ SOL(I,J): NEXT J
 90 NEXT I
100 CLS: PRINT TAB(5) "PROGRAM GASSOL": PRINT
110 PRINT "THIS PROGRAM PROVIDES SOLUBILITY
    VALUES FOR THE FOLLOWING GASES IN"
120 PRINT "WATER (GAS PRESSURE, INCLUSIVE OF
    AQUEOUS VAPOUR PRESSURE, 1 atm): "
130 PRINT: PRINT TAB(5) "1. OXYGEN"
140 PRINT TAB(5) "2. OZONE.
150 PRINT TAB(5) "3. METHANE "
160 PRINT TAB(5) "4. NITROGEN"
170 PRINT TAB(5) "5. AIR "
180 PRINT TAB(5) "6. CARBON DIOXIDE "
190 PRINT TAB(5) "7. CHLORINE "
200 PRINT TAB(5) "8. CHLORINE DIOXIDE "
210 PRINT TAB(5) "9. AMMONIA"
220 PRINT TAB(5) "10. HYDROGEN SULPHIDE "
230 PRINT TAB(5) "11. SULPHUR DIOXIDE "
```

```
240 PRINT INPUT "TO SELECT A GAS, ENTER ITS
    NUMBER:"; NUMB:1=NUMB
250 IF I=1 THEN GAS$="OXYGEN": IF I=2 THEN
    GAS$="OZONE": IF I=3 HEN GAS$="METHANE"
260 IF I=2 THEN GAS$="OZONE"
270 IF I=3 THEN GAS$="METHANE"
280 IF I=4 THEN GAS$="NITROGEN": IF I=5 THEN
    GAS$="AIR"
290 IF I=5 THEN GAS$="AIR"
300 IF I=6 THEN GAS$="CARBON-DIOXIDE":IF I=7
    THEN GAS$="CHLORINE"
310 IF I=7 THEN GAS$="CHLORINE"
320 IF I=8 THEN GAS$="CHLORINE DIOXIDE": IF I=9
    THEN GAS$="AMMONIA"
330 IF I=9 THEN GAS$="AMMONIA"
340 IF I=10 THEN GAS$="HYDROGEN SULPHIDE": IF
    I=11 THEN GAS$="SULPHURE DIOXIDE"
350 IF I=11 THEN GAS$="SULPHUR DIOXIDE"
360 IF I<6 THEN MAX=50 ELSE MAX=30
370 IF I<6 THAN UNIT$="mg/l" ELSE UNIT$="%wt."
380 PRINT: PRINT "ENTER WATER TEMPERATURE (deg
    C) IN RANGE 0-"; MAX
390 INPUT T
400 J=INT(T/5)+1
410 SATC=SOL(I,J)+(T-(J-1)*5)*(SOL(I,J+1)-
    SOL(I,J))/5
420 PRINT: PRINT "SOLUBILITY OF ";GAS$;"AT
    ";T;"deg C"
430 PRINT "IS ";SATC;" "; UNIT$
440 DATA 69.9, 60.9, 53.8, 48.1, 43.4, 39.4, 36.1
    33.2, 30.8, 28.5, 26.5
450 DATA 41.4, 35.8, 29.8, 24.1, 18.9, 13.7, 9.6,
    6.6, 4.3, 2.7, 1.6
460 DATA 39.8, 34.4, 30, 26.4, 23.7, 21.5, 19.9,
    18.3, 17.1, 16.1, 15.4
470 DATA 29.4, 26, 23.1, 20.8, 19, 17.5, 16.2, 15,
    13.9, 13, 12.2
480 DATA 38.89, 34.25, 30.49, 27.43, 24.89, 22.76,
    20.99, 19.44, 18.12, 16.90, 15.79
490 DATA 0.3346, 0.2274, 0.2318, 0.1970, 0.1688,
    0.1149, 0.1257
500 DATA 1.330, 1.160, 0.990, 0.850, 0.730, 0.640,
    0.567
510 DATA 19.08, 15.38, 12.46, 10.12, 8.25, 6.74,
    5.52
520 DATA 98.0, 81.0, 69.5, 60.0, 51.7, 45.2, 40.4
530 DATA 0.7066, 0.6001, 0.5112, 0.4411, 0.3846,
    0.3375, 0.2983
540 DATA 22.83, 19.31, 16.21, 13.54, 11.28, 9.41,
    7.80
```

LISTING OF THE PROGRAM DECANT.BAS:

```
10 REM PROGRAME DECANT: Analyses steady gradually
   varied flow in channels
20 PRINT: PRINT "Program DECANT": PRINT
30 PRINT "THIS PROGRAM ANALYSES STEADY
   GRADUALLY VARIED FLOW IN OPEN'
40 PRINT "CHANNELS WITH UNIFORM LATERAL INFLOW
   OVER THE CHANNEL LENGTH. "
50 PRINT "THE FLOW DEPTH AT THE OUTLET END IS
   ASSUMED TO BE CRITICAL DEPTH, "
  60 PRINT "SUCH AS WOULD BE THE CASE AT A FREE
     OVER-FALL. "
  65 PRINT "THE FRICTION HEAD LOSS IS COMPUTED
     USING the MANNING OR"
  66 PRINT "DARCY-WEISBACH FLOW EQUATIONS, AS
     SELECTED BY THE USER. "
  70 PRINT "THE PROGRAM USES A FOURTH-ORDER
     RUNGE-KUTTA NUMERICAL"
  80 PRINT "COMPUTATIONAL SCHEME IN THE
     SOLUTION OF THE WATER SURFACE SLOPE"
  90 PRINT "EQUATION (4,20), STARTING FROM THE
     OUTLET END WHERE the DEPTH"
 100 PRINT "IS TAKEN (FOR COMPUTATIONAL
     REASONS) TO BE 1.02 TIMES the CRITICAL
     DEPTH. "
 110 PRINT
 880 PRINT: PRINT "DATA ENTRY: "
 890 PRINT: PRINT "DO YOU WISH TO USE 1 MANNING
     OR 2 DARCY- WEISBACH ?"
 900 INPUT "ENTER 1 OR 2, AS APPROPRIATE"; NUMB
 910 IF NUMB =1 THEN FORM$="M": INPUT "MANNING
     N-VALUE"; MN
 920 IF NUMB =1 THEN FORM$="DW": INPUT "WALL
     ROUGHNESS (mm) "; KS:KS=KS/1000
 930 PRINT: PRINT "ENTER CHANNEL DATA: "
 940 PRINT "IS SECTION 1 CIRCULAR 2 RECTANGULAR
     3 TRPEZOIDAL ?"
 950 INPUT "ENTER 1, 2 OR 3, AS APPROPRIATE"; NO
 960 IF NO=1 THEN SECT$="CIRC": INPUT "DIAMETER
     (mm) "; D: D=D/1000
 970 IF NO=2 THEN SECT$="RECT": INPUT "CHANNEL
     WIDTH (mm) "; B: B=B/1000
 980 IF NO=3 THEN SECT$="TRAP": INPUT "BOTTOM
     WIDTH (mm) "; B: B=B/1000
 990 IF NO=4 THEN SECT$="ANGLE OF SIDE TO HORL
     (deg) "; FI
1000 PRINT: INPUT "ENTER CHANNEL BED SLOPE";
     FO
1010 PRINT: INPUT "ENTER DISCHARGE (m**3/s) ";
     Q0
```

```
1020 PRINT: INPUT "ENTER CHANNEL LENGTH (m)";
     XL
1030 PRINT: INPUT "ENTER CHANNEL STEP
     COMPUTATION LENGTH (m)": DELX
1040 PRINT: PRINT" .....data input complete;
     computation now in progress....."
1050 DIST=0: G=9.810001:DELX=-DELX:Q=QO:
     DISTT=DIST+XL
1060 QL=QO/XL
1070 REM******COMPUTER CRITICAL DEPTH*******
1080 GOSUB 2290
1090 PRINT: PRINT "CRITICAL DEPTH (mm) =
     ";YC*1000
1100 YO=1.02*YC: Y=YO
1110 PRINT: PRINT "DISTANCE X (m) DEPTH Y (mm) Q
     (m**3/s)"
1120 PRINT using "#########.##"; DISTT,
     YO*1000, QO
1130 IF FORM$="M"THEN GOSUB 2040
1140 IF FORM$="DW"THEN GOSUB 2080
1150 A1=(FO-FS-QL*Q/(G*A*A))/(1-Q*Q*TW/
     (G*A^3))
1160 Y=YO+5*A1*DELX:Q=Q+.5*DELX*QO/XL
1170 IF FORM$="M"THEN GOSUB 2040
1180 IF FORM$="DW"THEN GOSUB 2080
1190 A2=(FO-FS-QL*Q/(G*A*A))/(1-Q*Q*TW/
     (G*A^3))
1200 Y=YO+5*A2*DELX
1210 IF FORM$="M"THEN GOSUB 2040
1220 IF FORM$="DW"THEN GOSUB 2080
1230 A3=(FO-FS-QL*Q/(G*A*A))/(1-Q*Q*TW/
     (G*A^3))
1240 Y=YO+5*A3*DELX.Q=Q+.5*DELX*QO/XL
1250 IF FORM$="M"THEN GOSUB 2040
1260 IF FORM$="DW"THEN GOSUB 2080
1270 A4=(FO-FS-QL*1/(G*A*A))/(1-Q*Q*TW/
     (G*A^3))
1280 DELY=(DELX/6)*(A1+2*A2+2*a3+A4)
1290 Y=YO+DELY: YO=Y
1300 DIST=DIST+DELX:DISTT=DIST+XL
1310 PRINT USING "#########.##"; DISTT, Y*1000,
     Q
1320 IF ABS (DIST+DELX)XL THEN GOTO 1330 ELSE
     GOTO 1130
1330 PRINT: INPUT "DO YOU WISH TO MAKE ANOTHER
     COMPUTATION (Y/N)"; ANS$
1340 IF ANS$="Y"THEN GOTO 880
1350 END
2040 REM**SUBROUTINE TO COMPUTE FRICTION SLOPE
     USING MANNING EQN**
2050 GOSUB 2130
```

```
2060 FS=(MN*1/A)^2*(RH^-1.33)
2070 RETURN
2080 REM**SUBROUTINE TO COMPUTE FRICTION SLOPE
     USING WEISBACH EQN**
2090 GOSUB 2130
2100 V=ABS(Q/A): K=KS: GOSUB 2500
2110 FS=F*Q*Q/(8*9.810001*A*A*RH)
2120 RETURN
2130 REM **SUBROUTINE TO CALCULATE A, RH and
     TW**
2140 IF SECT$="CIRC" THEN GOTO 2180
2150 IF SECT$="RECT" THEN A=B*Y:RH=B*Y/
     (B+2*Y):TW=B: RETURN
2160 IF SECT$="TRAP" THEN A=Y*(B+Y/
     TAN(FI*3.142/180))
2170 RH=A/(B+(2*Y)/SIN(FI*3.142/180)):
     TW=B+(2*Y)/TAN(FI*3.142/180):RETURN
2180 HI=3.1416: LO=0!
2190 TH=(HI+LO)/2
2200 XR=1-(2*Y)/D-COS(TH)
2210 IF XR<0 THEN LO-TH
2220 IF XR>0 THEN HI=TH
2230 Z=(HI+LO)/2
2240 IF ABS(Z-TH)>.001 THEN GOTO 2190
2250 P=D*TH: A=.25*D*D(TH-.5*SIN(2*TH)): RH=
     A/P
2260 TW=D*SIN(TH)
2270 RETURN
2280 REM **SUBROUTINE TO COMPUTE CRITICAL
     DEPTH**
2290 IF SECT$="CIRC" THEN GOTO 2410
2300 IF SECT$="RECT" THEN YC=(Q^2/
     B^2*G))^.3333: RETURN
2310 IF SECT$="TRAP" THEN XC=Q^2/G
2320 HI=20!:LO=0!
2330 YY=(HI+LO)/2
2340 BB=YY/TAN(FI*3.142/180):
     A=(B+BB)*YY:BBB=B+2*BB
2350 XR=A^3/BBB-XC
2360 IF XR<0 THEN LO=YY
2370 IF XR>0 THEN HI=YY
2380 Z=(HI+LO)/2
2390 IF ABS(Z-YY)>.001 THEN GOTO 2330
2400 YC=YY: RETURN
2410 HI=3.142: LO=0!
2420 TH=(HI+LO)/2
2430 A=.25*D*D*(TH-
     .5*SIN(2*TH)):BBB=D*SIN(TH)
2440 XR=A^3/BBB-Q^2/G
2450 IF XR<0 THEN LO=TH
2460 IF XR>0 THEN HI=TH
```

```
2470 Z=(HI+LO)/2
2480 IF ABS(Z-TH).001 THEN GOTO 2420
2490 YC=.5*D*(1-COS(TH)):RETURN
2500 REM **SUBROUTINE FRICTION FACTOR**
2510 KVISC=1.307E-06
2520 UPV=.5
2530 LOV=0
2540 F=(UPV+LOV)/2
2550 YY=1/SQR(F)
2560 X=K/(14.8*RH)+(2.51*KVISC)/
     (4*RH*V*SQR(F))
2570 W=YY+88*LOG(X)
2580 IF W<0 THEN UPV=F
2590 IF W>0 THEN LOV=F
2600 Z=(UPV+LOV)/2
2610 E=ABS((Z-F)/F)
2620 IF E<.005 THEN GOTO 2640
2630 F=Z: GOTO 2550
2640 F=Z
2650 RETURN
2660 REM **SUBROUTINE TO COMPUTE NORMAL DEPTH**
2670 FSR=FO^.5
2680 HI=40:LO=.001
2690 IF SECT$="CIRC" THEN GOTO 2800
2700 IF SECT$="RECT" THEN GOTO 2910
2710 IF SECT$="TRAP" THEN GOTO 2720
2720 Y=(HI+LO)/2
2730 A=Y*(B+Y/TAN(FI*3.142/180)): RH=A/
     (B+(2*Y)/SIN(FI*3.142/180))
2740 IF FORM$="M" THEN FSH=MN*Q/(A*RH^.67):
     GOTO 2770
2750 IF FORM$="DW" THEN V=Q/A:K=KS:GOSUB 2500
2760 FSH=(F/(8*9.810001*RH))^.5*V
2770 GOSUB 2990
2780 IF ABS(Z-Y)>.0002 THEN GOTO 2720
2790 RETURN
2800 HI=3.1416:LO=.001
2810 TH=(HI+LO)/2
2820 A=.25*D*D*(TH-.5*SIN(2*TH))
2830 P=D*TH:RH=A/P
2840 IF FORM$="M" THEN FSH =MN*Q/
     (A*RH^.67):GOTO 2870
2850 IF FORMS$="DW" THEN V=Q/A:K=KS:GOSUB
     2500
2860 FSH=(F/(8*9.810001*RH)^.5*V
2870 GOSUB 3050
2880 IF ABS(FSH-FSR)/FSR.001 THEN GOTO 2810
2890 Y=.5*D*(1.COS(TH))
2900 RETURN
2910 Y=(HI+LO)/2
2920 A=B*Y:RH=A/(B+2*Y)
```

```
2930 IF FORM="M" THEN FSH=MN*Q/(A*RH^.67):GOTO
     2960
2940 IF FORMS="DW" THEN V=Q/A:K=KS:GOSUB 2500
2950 FSH=(F/(8*9.810001*RH))^.5*V
2960 GOSUB 2990
2970 IF ABS(Z-Y)>.0002 THEN GOTO 2910
2980 RETURN
2990 REM INTERVAL-HALFING ROUTINE
3000 WW-FSH-FSR
3010 IF WW>0 THEN LO=Y
3020 IF WW<0 THEN HI=Y
3030 Z=(HI+LO)/2
3040 RETURN
3050 WW=FSH-FSR
3060 IF WW>0 THEN LO=TH
3070 IF WW<0 THEN HI=TH
3080 RETURN
3090 REM CALC OF WEIR DISCHARGE RATE QW
3100 CW=4.15-1.81*YC/Y-.14*YC/LW.HW=Y-
     (HC+DIST*FO)
3110 IF HW<0 THEN PRINT "WATER LEVEL BELOW WEIR
     CREST"GOTO 3130
3120 QW=NW*CW*HW^1.5
3130 RETURN
```

LISTING OF THE PROGRAM OCFIT.BAS:

```
 10 CLS
 20 REM PROGRAM OCFIT
 30 =======================================
 40 PRINT "COMPUTATION OF THE OC OF DIFFUSED AIR
    AERATION SYSTEMS"
 50 PRINT
 60 PRINT "DATA SOURCE MAY BE: disk file"
 80 PRINT "keyboard input"
100 PRINT
110 PRINT "COMPUTATION PROCEDURE: by curve
    fitting as set out in the"
120 PRINT "ASCE Manual for OC Determination.
130 PRINT
140 PRINT "COMPUTED PARAMETERS: kla.OC.%OU, SAE
150 '=======================================
160 PRINT:PRINT "PRESS THE SPACE BAR TO
    CONTINUE"
170 DIM C(150), T(150), LD(150), CC(150,4),
    CT(150), R(150), FIT(150,4), ACO(4), ACS(4),
    F(150), KLA(10), OC(10)
180 ST=0:SDO=0:T2=0:TDO=0;SR=0:SS=0:SUMD=0:
    MIND=1000:PRINT
190 Y$=INPUT$(1)
```

```
200 CLS
220 INPUT "Do you wish to retrieve test data from
    disk (Y/N)";ANS$
230 IF ANS$="Y" THEN INPUT "Enter data file name";
    A$:GOSUB 3630: GOTO 420
235 REM INPUT OF DATA FROM KEYBOARD
236 INPUT "Enter STP name (in ICs)";STP$
237 INPUT "Enter date of test (in ICs)" ;CAL$
238 INPUT "Enter test reference (data file
    name)";A$
242 PRINT "SYSTEM TYPE:1. Surface aeration system
    "2. Diffused air system"
244 PRINT "2. Diffused air system"
246 INPUT "Enter 1 or 2, as appropriate"; NUMB
248 IF NUMB=1 THEN DS=0:GOTO 260
250 INPUT "Enter Diffuser Submergence (m)";DS
260 PRINT:INPUT "Enter water volume in test tank
    (m^3)";VW
270 INPUT "Enter water temperature (deg C)";TW
300 GOSUB 3240
390 PRINT "Computation in progress; please wait"
420 REM COMPUTATION OF KLA BY LEAST SQUARES
430 DD=0:T(0)=-INTV:ND=NPSET
440 FOR J=1 TO NSET
450 FOR I=1 TO ND.C(I)=CC(I,J):T(I)=T(I-
    1)+INTV:NEXT I
460 IF C(ND)=0 THEN KLA(J)=0:DD=DD+1:GOTO 860
470 CS=100+(DS*4.5): CO=C(I) :KLA=LOG((CS-CO)/
    (CS-C(ND)))/(T(ND)- T(1)) 480 NC=0
490 OS=0
500 FOR 1=1 TO ND
510 F(I) =CS-(CS-CO)*EXP(-KLA*T(I))
520 R(I)= C(I) -F(I)
530 OS=OS+R(I)*R(I)
540 NEXT I
550 NC=NC+1
560 A1=0:A2=0:A3=0:A4=0:A5=0:A6=0:C1=0:
    C2=0:C3=0:SQ=0
570 REM SETUP NORMAL EQUATIONS FOR LINEARISED
    MODEL
580 REM USING CURRENT LEAST SQUARES ESTIMATES
    590 FOR I=1 TO ND
600 Z2=EXP(-KLA*T(I)):ZI=1-Z2: Z3=T(I)*Z2*
    (CS-CO)
610 A1=A1+Z1*Z1: A2=A2+Z1*Z2: A3=A3+Z1*Z3:
    A4=A4+Z2*Z2: A5=A5+Z2*Z3: A6=A6+Z3*Z3
620 F(I) =CS-(CS-CO)*Z2
630 R(I)=C(I) -F(I)
640 C1=C1+R(I)*ZI: C2=C2+R(I)*Z2:
    C3=C3+R(I)*Z3
650 NEXT I
```

```
660 REM SOLUTION OF NORMAL EQUATIONS FOR
    CORRECTIONS
670 REM TO THE PRIOR LEAST SQUARES ESTIMATES
680 D1=A1*A4-A2*A2:D2=A1*C3-A3*C1:D3=A1*A5-
    A3*A2:D4=A6*A1- A3*A3:D5=A1*C2-A2*C1
690 XN=D1*D2-D3*D5: XD=DI*D4-D3*D3: X3=XN/
    XD:YN=D5-D3*X3: X2=YN/D1
700 X1=(C1-A2*X2-A3*X3)/A1
710 REM UPDATE ESTIMATES, SUM OF SQUARES
720 T1=X1+C2: T2=X2+C0: T3=X3+KLA
730 FOR I=1 TO ND
740 F(I)=T1-(T1-T2)*EXP(-T3*T(I))
750 R(I)-C(I) -F(I)
760 SQ=SQ+R(I)*R(I)
770 NEXT I
780 REM TEST FOR CONVERGENCE
790 IF(X1/T1<=.00001) AND (X2/T2<=.00001) AND
    (X3/T3.00001) GOTO 840
800 IF(ABS(OS-SQ)<=.000001) GOTO 840
810 IF (NC>10) THEN PRINT "SOLUTION NOT
    CONVERGED IN TEN ITERS. "GOTO 1920
820 CS=T1: C0=T2: KLA=T3: OS=SQ
830 GOTO 550
840 CS=T1: CO=T2: KLA=T3: OS=SQ
850 KLA(J)=KLA:ACS(J)=CS:ACO(J)=C0
860 NEXT J
870 F=1.024:CS20=9.08: OCAVE=0.KLAVE=0
875 REM COMPUTATION OF OXYGENATION CAPACITY
    (OC) KG/H 880- FOR J- 1 TO NSET
940 OC(J)=KLA(J)*F^(20-TW)*(ACS(J)/
    100)*CS20*VW*60/1000 945
    OCAVE=OCAVE+OC(J)
946 KLAVE=KLAVE-KLA(J)
950 NEXT J
955 OCAVE=OCAVE/NSET:KLAVE=KLAVE/NSET
1000 REM COMPUTATION OF PERCENT OXYGEN UPTAKE
     (POU)
1010 POU=100*OCAVE/(QM*.2314)
1020 REM COMPUTATION OF STANDARD AERATION
     EFFICIENCY (SAE)
1040 SAE=OCAVE/POW
1050 REM PRINT RESULTS
1070 FOR J=1 TO NSET
1080 FOR I=1 TO ND
1090 FIT(I,J)=ACS(J)-(ACS(J)-ACO(J))*EXP(-
     KLA(J)*T(I)) 1100 NEXT I
1110 NEXT J
1120 CLS
1130 PRINT"COMPARISON OF MEASURED (M) AND
     FITTED (F) DO VALUES ":PRINT
1140 PRINT" TIME M1 F1 M2 F2 M3 F3 M4 F4 "
```

```
1150 PRINT" (MIN.) "
1160 FOR I=1 TO ND
1170 PRINT USING "###.# "; T(I), CC(I,1),
     FIT(I,1), CC(1,2), FIT(1,2),
     CC(1,3),FIT(1,3),CC(1,4), FIT(1,4)
1180 NEXT I
1190 PRINT:PRINT "TO CONTINUE PRESS A KEY"
1200 X$=INPUT$(I)
1210 PRINT"RESULTS FOR TEST
     REFERENCE ":A$:PRINT
1215 IF NUMB=1 THEN GOTO 1260
1220 PRINT"Air mass inflow rate (kg/h) =";QM
1260 PRINT
1280 FOR J=1 TO NX
1290 PRINT"Computed KLA";J;KLA(J)
1300 NEXT J
1330 PRINT
1340 PRINT"Standard mean oxygenation capacity (kg
     O2/h) =";OCAVE
1345 IF NUMB=1 THEN GOTO 1360
1350 PRINT"Percent oxygen absorption =';POU
1360 PRINT
1370 PRINT"Power input (kW) =";POW
1380 PRINT"Standard aeration efficiency (kg O2/
     kWh) =";SAE
1390 PRINT:INPUT "Do you wish to print results on
     paper (Y/N) ";ANS$
1400 IF ANS$="Y" THEN GOTO 1410 ELSE GOTO 1880
1410 REM PRINTING RESULTS ON PAPER
1420 LPRINT:LPRINT:LPRINT:LPRINT
1430 LPRINT TAB(12) STP$:LPRINT
1440 LPRINT TAB(12) "OC TEST CARRIED OUT
     ON";CAL$
1450 LPRINT TAB(12) "RESULTS FOR TEST
     REF.",A$:LPRINT:LPRINT
1460 LPRINT TAB(12) "RE-AERATION DO
     MEASUREMENTS":LPRINT
1470 LPRINT TAB(12) "TIME DO 1 DO 2 DO 3 DO 4"
1475 LPRINT TAB(12) "(MIN) %SAT %SAT %SAT
     %SAT":LPRINT
1570 FOR I=1 TO ND
1580 LPRINT TAB(12) USING "###.# ";T(I),
     CC(I,1), CC(I,2), CC(I,3), CC(I,4)
1590 NEXT I
1690 LPRINT:LPRINT:LPRINT
1700 LPRINT TAB(12) "Water Temperature (deg C)
     =";TW
1710 LPRINT TAB(12) "Water Volume (m^3) =";VW
1715 IF NUMB-1 THEN GOTO 1730
1720 LPRINT TAB(12) "Air flow rate (kg/h) =";QM
1730 LPRINT
```

```
1750 FOR J=1 TO NSET
1760 LPRINT TAB(12) "Computed KLA"; J; KLA(J)
1761 NEXT J
1782 LPRINT
1790 LPRINT TAB(12) "Mean value of KLA (1/min)
     =";KLAVE
1820 LPRINT
1830 LPRINT TAB(12) "Standard mean OC value (kg
     O2/h) =";OCAVE
1835 LPRINT
1836 IF NUMB=1 THEN GOTO 1845
1837 LPRINT TAB(12) "Percent oxygen uptake =";POU
1840 LPRINT
1845 LPRINT TAB(12) "Standard aeration efficiency
     (kg O2/kWh) =";SAE
1850 LPRINT
1880 PRINT:INPUT "Do you wish to store the date on
     disk (Y/N)";ANS$
1890 IF ANS$="Y" THEN GOSUB 3520
1900 PRINT:INPUT "Do you wish to enter another set
     of data (Y/N)":ANS$
1910 IF ANS$="Y" THEN GOTO 200
1920 END
3240 REM subroutine to enter data by the keyboard
3250 INPUT "ENTER SYSTEM POWER INPUT (kW) ";POW
3255 IF NUMB-2 THEN INPUT "ENTER MASS AIR FLOW
     RATE (kg/h) ";QM
3310 INPUT "ENTER TIME INTERVAL BETWEEN READINGS
     (MIN) ";INTV
3320 INPUT "HOW MANY DO DATA SETS DO YOU WISH TO
     ENTER ";NSET
3330 INPUT "HOW MANY DO VALUES IN EACH SET ";NPSET
3340 T(I)=0
3350 FOR 1=2 TO NPSET:T(1)=T(I-1)+INTV:NEXT I
3360 PRINT
3370 FOR J-1 TO NSET
3380 PRINT "ENTER VALUES FOR DATA SET ";J
3390 FOR I=1 TO NPSET
3400 INPUT CC(I,J); NEXT I
3410 PRINT: NEXT J
3420 CLS
3430 PRINT "DATA CHECK; TIME (MIN), DO VALUES (%
     SAT)"
3440 PRINT "TIME DO 1 DO 2 DO 3 DO 4":J=1
3450 FOR I=1 TO NPSET
3460 PRINT USING "#### ";T(I);
3470 PRINT USING "####.#";
     CC(I,J),CC(I,J+1),CC(I,J+2),CC(I,J+3)
3480 J=1
3490 NEXT I
3500 ND=NPSET
```

```
3510 RETURN
3520 REM TRANSFER OF DATA TO DISK FILE
3530 OPEN "O",1,A$
3540 WRITE# 1,INTV,NSET,NPSET,VW,TW,QM,POW,
     DS,STP$,CAL$
3550 FOR I=1 TO NPSET
3560 FOR J=1 TO NSET
3570 WRITE# 1.CC(I,J)
3580 NEXT J
3590 NEXT I
3600 CLOSE I
3610 RETURN
3620 REM RETRIEVAL OF DATA FROM DISK FILE
3630 OPEN "I",1.A$
3640 INPUT# 1,INTV,NSET.NPSET,VW,TW.QM,POW,
     DS,STP$,CAL$
3650 FOR I=1 TO NPSET
3660 FOR J=1 TO NSET
3670 INPUT# 1.CC(I,J)
3680 NEXT J
3690 NEXT I
3700 CLOSE I
3710 RETURN
```

Appendix B
ATOMIC WEIGHTS OF THE ELEMENTS (amu)

From WATER CHEMISTRY, Snoeyink & Jenkins, Wiley 1980

	Symbol	Atomic number	amu		Symbol	Atomic number	amu
Actinium	Ac	89		Gadolinium	Gd	64	157.25
Aluminium	Al	13	26.98	Gallium	Ga	31	69.72
Americum	Am	95		Germanium	Ge	32	72.59
Antimony	Sb	51	121.75	Gold	Au	79	196.97
Argon	Ar	18	39.95	Hafnium	Hf	72	178.49
Arsenic	As	33	74.92	Helium	He	2	4.00
Astatine	At	85		Holmium	Ho	67	164.93
Barium	Ba	56	137.34	Hydrogen	H	1	1.01
Berkelium	Bk	97		Indium	In	49	114.82
Beryllium	Be	4	9.01	Iodine	I	53	126.90
Bismuth	Bi	83	208.98	Iridium	Ir	77	192.22
Boron	B	5	10.81	Iron	Fe	26	55.85
Bromine	Br	35	79.90	Krypton	Kr	36	83.80
Cadmium	Cd	48	112.4	Lanthanium	La	57	138.91
Calcium	Ca	20	40.08	Lawrencium	Lr	103	
Californium	Cf	98		Lead	Pb	82	207.20
Carbon	C	6	12.01	Lithium	Li	3	6.94
Cerium	Ce	58	140.12	Lutetium	Lu	71	174.97
Cesium	Cs	55	132.91	Magnesium	Mg	12	24.31
Chlorine	Cl	17	35.45	Manganese	Mn	25	54.94
Chromium	Cr	24	52.00	Mendelevium	Md	101	
Cobalt	Co	27	58.93	Mercury	Hg	80	200.59
Copper	Cu	29	63.54	Molybdenum	Mo	42	95.94
Curium	Cm	96		Neodymium	Nd	60	144.24
Dysprosium	Dy	66	162.50	Neon	Ne	10	20.18
Einsteinium	Es	99		Neptunium	Np	93	237.05
Erbium	Er	68	167.26	Nickel	Ni	28	58.71
Europpium	Eu	63	151.96	Niobium	Nb	41	92.91
Fermium	Fm	100		Nitrogen	N	7	14.01
Fluorine	F	9	19.00	Nobelium	No	102	
Francium	Fr	87		Osmium	Os	76	190.2

	Symbol	Atomic number	amu		Symbol	Atomic number	amu
Oxygen	O	8	16.00	Sodium	Na	11	22.99
Palladium	Pd	46	106.40	Strontium	Sr	38	87.62
Phosphorus	P	15	30.97	Sulphur	S	16	32.06
Platinum	Pt	78	195.09	Tantalum	Ta	73	180.95
Plutonium	Pu	94		Technetium	Tc	43	98.91
Polonium	Po	84		Tellurium	Te	52	127.60
Potassium	K	19	39.10	Terbium	Tb	65	158.93
Praseodymium	Pr	59	140.91	Thallium	Tl	81	204.37
Promethium	Pm	61		Thorium	Th	90	232.04
Protactinium	Pa	91	231.04	Thullium	Tm	69	168.93
Radium	Ra	88	226.02	Tin	Sn	50	118.69
Radon	Rn	86		Titanium	Ti	22	47.90
Rhenium	Re	75	186.21	Tungsten	W	74	183.85
Rhodium	Rh	45	102.91	Uranium	U	92	238.03
Rubidium	Rb	37	85.47	Vanadium	V	23	50.94
Ruthenium	Ru	44	101.07	Xenon	Xe	54	131.30
Samarium	Sm	62	150.40	Ytterbium	Yb	70	173.04
Scandium	Sc	21	44.96	Yttrium	Y	39	88.91
Selenium	Se	34	78.96	Zinc	Zn	30	65.38
Silicon	Si	14	28.09	Zirconium	Zr	40	91.22
Silver	Ag	47	107.87				

Appendix C
Computation of Oxygen Transfer Parameters

C.1 AERATION SYSTEMS

The types of aeration system used in the activated sludge process of wastewater treatment are discussed in Chapter 14. Essentially they can be categorised into two groups: (a) surface aeration systems and (b) diffused air systems. The performance of these systems is conventionally expressed in terms of a standard oxygenation capacity (OC), which is the rate that the system transfers oxygen to clean water at a water temperature of 20°C, at zero dissolved oxygen concentration and at an atmospheric pressure of 1 atm. As outlined in Chapter 14, the OC of an aeration system is measured *in situ* by first deoxygenating the system water and then monitoring its re-oxygenation rate, which can be modelled as follows (equation (14.21)

$$C_t = C_* - (C_* - C_i)\exp(-k_L a \cdot t) \tag{14.21}$$

where C_* is the oxygen saturation concentration for the aeration system, C_i is the initial oxygen concentration k_{La} is the overall transfer coefficient (t^{-1}) and t is the aeration time.

Program OCFIT.BAS analyses the test experimental oxygen concentration/time (C_t/t) data from an OC determination test. It uses regression analysis to find the best-fitting values for the oxygen transfer parameters c_i, k_{La} and C_* (ASCE, 1984). It is desirable that the data should extend over a wide span of oxygen concentration, e.g. in the range 15–95% saturation. Program OCFIT has been adapted from a program presented in the ASCE Standard, to which the reader is referred for a comprehensive discussion of oxygen transfer measurement.

Program OCFIT (program listing in Appendix A) has been written to cater for data generated by a set of DO meters, calibrated to read 100% (C_{sm}) in water at equilibrium with atmospheric air. It should be noted that the oxygen saturation concentration C_* for a diffused air system will exceed C_{sm} by an amount which is dependent upon the submergence of the diffuser. Program OCFIT takes this into account when used to analyse test data from diffused air systems. It also caters for surface aeration systems where $C_{sm} \cong C^*$

C.2   SAMPLE PROGRAM RUN: PROGRAM OCFIT

Run
        PROGRAM OCFIT

Computation of the OC of aeration systems

Data source may be: disk file
                          keyboard input

Computation procedure: by curve-fitting by procedure set out in the ASCE Standard (1985)

Computed parameters: kLa, OC, %OU, SAE

PRESS THE SPACE BAR TO CONTINUE

Do you wish to retrieve test data from disk (Y N)? Y

Enter file name ? VARRA_1

COMPUTATION IN PROGRESS, PLEASE WAIT

COMPARISON OF MEASURED (M) AND FITTED (F) DO VALUES

Time (min)	M1	F1	M2	F2	M3	F3	M4	F4
0	39.0	37.1	40.0	37.7	40.0	37.5	39.0	36.7
2	47.0	47.4	47.0	47.9	47.0	48.2	46.0	47.3
4	55.5	56.4	55.0	56.7	56.0	57.6	55.0	56.5
6	63.0	64.3	63.0	64.5	65.0	65.8	64.0	64.5
8	71.0	71.1	71.0	71.3	72.0	73.0	71.0	71.6
10	76.0	77.1	77.0	77.2	79.0	79.3	78.0	77.8
12	83.0	82.3	83.0	82.4	85.0	84.8	83.0	83.2
14	86.0	86.8	87.0	86.9	90.0	89.6	87.0	88.0
16	92.0	90.8	92.0	90.8	94.0	93.7	93.0	92.1
18	95.0	94.2	95.0	94.3	99.0	97.4	97.0	95.8
20	97.0	97.2	97.0	97.3	101.0	100.6	100.0	98.9
22	102.0	99.8	101.0	99.9	104.0	103.4	103.0	101.7
24	103.0	102.1	103.0	102.2	107.0	105.9	105.0	104.2
26	104.0	104.1	104.0	104.3	108.0	108.0	106.0	106.3
28	106.0	105.8	106.0	106.0	110.0	109.9	107.0	108.2
30	108.0	107.3	108.0	107.6	112.0	111.6	110.0	109.8
32	107.0	108.6	108.0	108.9	112.0	113.0	111.0	111.3
34	109.0	109.8	109.9	110.1	113.0	114.3	112.0	112.5
36	110.0	110.8	111.0	111.1	115.0	115.4	113.0	113.6

```
TO CONTINUE PRESS ANY KEY
RESULTS FOR TEST REFERENCE VARRA_1
```

Water volume (m³)	1518
Water temperature (°C)	5.6
Air mass flow rate (kg h⁻¹)	1520
Power input (kW)	35.8
Computed mean $k_L a$ value (1 min⁻¹)	0.067
Standard mean oxygenation capacity (kg O₂ h⁻¹)	93.21
Percent oxygen absorption	26.5
Standard aeration efficiency (kg O₂ kWh⁻¹)	2.6

```
Do you wish to print the results on paper (Y/N)? N
Do you wish to store the system data on disk (Y/N) N
Do you wish to enter another set of data (Y/N) N
Ok
```

Notes

1. The above data set was retrieved from disk. It relates to a fine bubble diffused air sysem in a 4 m deep tank. The program also allows direct input from the keyboard and may be adapted for data transfer from a data logger.

2. The procedure for the calculation of the *OC*-value of an aeration system is presented in Chapter 14.

3. The Fixed parameter values used in the programe are:
   Oxygen saturation concentration in air-equilibriated water at $20°C = 9.08$ mg l^{-1}
   $k_L a$ a temperature correction factor $f = 1.024$
   Oxygen mass fraction in air $= 0.23$

4. It is usual in carying out experimental oxygenation capacity tests to use a number of DO meters placed at different locations in the water volume being aerated (in the above example four DO meters were used). The program computes an *OC*-value for each meter and then calculates their arithmetic mean: hence the description standard mean *OC*-value. The program is written to cater for up to four data sets but can be easily adapted to cater for any larger number.

REFERENCE

American Society of Civil Engineers (1984) *ASCE Standard: Measurement of Oxygen Transfer in Clean Water*, New York.

INDEX